Power Systems

Electrical power has been the technological foundation of industrial societies for many years. Although the systems designed to provide and apply electrical energy have reached a high degree of maturity, unforeseen problems are constantly encountered, necessitating the design of more efficient and reliable systems based on novel technologies. The book series Power Systems is aimed at providing detailed, accurate and sound technical information about these new developments in electrical power engineering. It includes topics on power generation, storage and transmission as well as electrical machines. The monographs and advanced textbooks in this series address researchers, lecturers, industrial engineers and senior students in electrical engineering.

Power Systems is indexed in Scopus

More information about this series at http://www.springer.com/series/4622

Vasily Ya. Ushakov · Alexey V. Mytnikov ·
Valeriy A. Lavrinovich · Alexey V. Lavrinovich

Transformer Condition Control

Advanced and Traditional Technologies

 Springer

Vasily Ya. Ushakov
Tomsk Polytechnik University
Tomsk, Russia

Valeriy A. Lavrinovich
Russian Federal Nuclear Center
Russia Research Institute of Technical
Physics
Moscow, Russia

Alexey V. Mytnikov
Institute of Power Engineering
Tomsk Polytechnic University
Tomsk, Russia

Alexey V. Lavrinovich
Tomsk Oil and Gas Research and Design
Institute
Tomsk, Russia

Translated by
Pavel Yu. Moshin
Tomsk State University
Tomsk, Russia

ISSN 1612-1287   ISSN 1860-4676 (electronic)
Power Systems
ISBN 978-3-030-83200-1   ISBN 978-3-030-83198-1 (eBook)
https://doi.org/10.1007/978-3-030-83198-1

This Springer imprint is published by the registered company Springer Nature Switzerland AG
The registered company address is: Gewerbestrasse 11, 6330 Cham, Switzerland

Preface

The United Nations (UN), the International Energy Agency (IEA), the European Energy Charter, and other supranational and national structures related to the problem of meeting a growing need of mankind for electrical energy, name among the most serious challenges the aging of the entire power infrastructure and, primarily, the aging of electric power grids (EPG). Power transformers are one of the most complex in design, expensive, and widespread elements of EPG. For example, the total number of power transformers in all voltage classes (except instrument transformers) as a whole across the Unified National Power Grid (UNPG) and Distribution Grid Companies (DGC) of Russia amounted as of 2011 to 643.3 thousand units, their installed capacity being 832.6 thousand MVA. The total capacity of power transformers in Russian power systems of all voltage classes above 3 kV is currently 6–6.5 times higher than the installed capacity of all power plant generators due to a technological need for several stages of voltage transformation in the process of electricity transmission and distribution. The main part of transformers (in terms of quantity) consists of transformers with voltage up to 20 kV (94.7%), which are installed most entirely in distribution networks that serve agricultural purposes. The UNPG facilities account for about 0.8% of transformers in terms of quantity, and for about 37.8% in terms of capacity, while the DGC facilities account for 99.2% of transformers in terms of quantity and 62.2% in terms of capacity. Of the total number of power transformers installed in UNPG networks with voltage of 110 kV and higher, 55.3% of the units have exhausted their resource (as of 01.01.2012), with the accepted standard resource of service life for basic electrotechnical equipment being 25 years.

Approximately the same situation is typical for all countries with a developed electric power industry. As in Russia, despite the fact that many transformers have completed their standard service life, they continue to be operated due to the reason of saving the finances needed for replacement. In order to prevent severe accidents due to transformer failures in electric power systems, regulatory documentation has been developed and is implemented to control the diagnostics of power transformer conditions. The regulatory documentation of the Russian UNPG, put into effect in 2017, prescribes complex diagnostic tests (CDT) of transformers according to 12 points:

- Chromatographic analysis of dissolved gases.
- Moisture content assessment of solid insulation.
- Insulation resistance measurement.
- Measurement of the tangent of the angle of dielectric losses in the insulation of the windings.
- Assessment of paper insulation conditions in windings.
- Insulation test using increased voltage of 50 Hz.
- Measurement of winding resistance to direct current.
- Checking the transformation ratio.
- Checking the group of connections in windings.
- Transformer phasing.
- Measurement of no-load losses at low voltage.
- Measurement of short-circuit resistance ($Z_{s.c}$).

Besides, if any additional data on transformer conditions are needed, then, depending on the decision of an enterprise, additional tests are made:

- Thermal image monitoring of transformer conditions.
- Measurement of partial discharge characteristics.

Despite such comprehensive CDT, sudden transformer failures continue to occur directly after an inspection issuing an act granting further operation. The reason for this to happen is the presence of hidden defects undetectable by regulatory diagnostic procedures. These defects include, for instance, displacement of winding coils relative to each other in different directions: axial and radial shifts, coil lowering, and other displacements. Any shift leads to an increase in the electric field strength between the shifted coils, and then to a subsequent speedup of insulation degradation in this locality, eventually causing a breakdown.

Transformers having mechanical displacements in their windings, such as coil displacement and buckling, or such as winding unpressurization, can be operated for a certain amount of time and cannot be diagnosed by any means available. Depending on the degree of alteration in the mechanical condition of windings, as compared with their standard condition, the process of insulation degradation may take long years of operation. Defects, however, can be so rapid in development that critical damage to a transformer may occur within just a few days after CDT procedures have been completed. Most often this takes place after a short circuit (SC) in the outgoing line (behind the transformer).

According to the Russian joint stock company Unified Energy System of Russia (RAO UES), transformers and autotransformers with voltage of 110...500 kV and capacity of 63 MVA and above, operated at enterprises of electrical and intersystem networks, show around 30% of the total number of technological failures as related to a disconnection of equipment by protection devices or personnel on an emergency request precisely due to SC occurrence inside the transformers. The main reasons for this type of SC are: wear and tear of winding insulation, insufficient electrodynamic firmness of windings against SC currents, internal insulation breakdowns of high-voltage bushings, damage to on-load voltage regulators (OLVR). Thus, two main causes of damage could be distinguished: internal insulation degradation and insufficient mechanical endurance of windings against SC currents.

The physical methods and technical means available for monitoring transformer conditions have a common drawback: they permit detecting only certain types of defects, and those only at advanced stages of development, as well as with significant errors involved. Research and operating experience show that available transformer control technologies reveal no more than 10% of existing defects. On the other hand, unreasonable rejection of fully functional equipment occurs unacceptably often. Unresolved issues of the sensitivity and reliability of methods and tools for diagnostics of high-voltage equipment, including transformers in particular, provide a standing subject of discussions at regular scientific and technical conferences and meetings of various levels, including the international level. Thus, there is every reason to state the absence of any standardized procedure for a proved and reliable monitoring of the condition of transformer active parts that meets the requirements for power systems in the twenty-first century.

The present monograph reflects the general state of the art in reliable and high-quality diagnostics of the active parts of transformers and suggests some means of solution to this problem. The most attention is paid to monitoring the condition of transformer windings by using the pulse method, since it has shown fairly high efficiency. In addition, the authors of the monograph have rather recently proposed and patented an improved version of this method, which is either unknown to non-Russian-speaking experts, or is not available in a form providing sufficient data for its practical implementation.

The monograph has been prepared for publication due to financial support granted by Prof. A. A. Yakovlev, Rector of TPU (National Research Tomsk Polytechnic University), and Prof. M. S. Yusubov, Vice-Rector for Research. The English translation has been made by Dr. P. Yu. Moshin. The authors express their sincere gratitude for this assistance.

Readers may send their comments and suggestions to the authors by e-mail at vyush@tpu.ru and mytnikov66@mail.ru.

Tomsk, Russia	Vasily Ya. Ushakov
Tomsk, Russia	Alexey V. Mytnikov
Moscow, Russia	Valeriy A. Lavrinovich
Tomsk, Russia	Alexey V. Lavrinovich

About This Book

This monograph is devoted to one of the main problem of modern electrical power engineering—power transformer diagnostics. The monograph consists of a preface, six chapters.

The preface substantiates the importance of the problem of diagnostics the state of power transformers for the reliable operation of electric power systems and, consequently, for uninterrupted power supply to consumers. It also expresses gratitude to those who contributed to the publication of this monograph.

The Chap. 1 contains materials about the causes of failure of power transformers, statistical distributions of failures of transformer design elements, technical and economic consequences of disruption of the normal operation of transformers.

The Chap. 2 describes the standardized (traditional) technologies for monitoring the state of the power transformers.

The Chap. 3 deals with a relatively new method for diagnosing the condition of transformers, based on sensing their windings with low-voltage pulses.

The Chap. 4 presents the results of research and development technologies for transformer windings condition control by means of nanosecond pulses, which were carried out by the authors of this monograph. The comparative efficiency analysis of the technologies of frequency analysis and pulse defectography is presented.

The appendixes contain the stages of development of a schematic diagram and a prototype of a probe pulse generator, and a program for testing and processing of diagnostic results.

Each chapter ends with a list of references.

Contents

Chapter 1
Causes of Power Transformer Failure

A high-voltage transformer is a complex device in both design and materials. Accordingly, the causes and effects of failures affecting both the structure elements of a transformer and the entire apparatus are quite diverse. Figure 1.1 shows the most common defects occurring in three transformer elements being the main assembly units.

Monitoring the condition of high-voltage power transformers, being expensive, critical and long-term elements of power systems, is based on registering alterations in the physicochemical and mechanical properties of materials, as well as in the geometric parameters of assembly units and structure elements, which are ultimately responsible for transformer failures. The origin of failures lies in electrical, thermal, mechanical and other influences on the normal operating mode, as well as in emergency situations, see Fig. 1.2.

1.1 Failure Statistics of Transformer Construction Elements

To determine the reasons for transformer failures, a study of incidents affecting two groups of transformers from different series and classes of voltage and rated power has been carried out. One of the groups of transformers under study included three-phase three-winding oil-filled power transformers with forced air circulation and natural oil circulation, transformers with voltage regulation under load, as well as three-phase two-winding oil power transformers of 110/6 kV with low-voltage split winding, forced air circulation, natural oil circulation, and voltage regulation under load. Analysis of incidents affecting the mentioned series of 110 kV voltage transformers with a rated power of 16,000–40,000 kVA makes it possible to state the reasons for transformer failures as listed below.

© The Author(s), under exclusive license to Springer Nature Switzerland AG 2022
V. Ya. Ushakov et al., *Transformer Condition Control*, Power Systems,
https://doi.org/10.1007/978-3-030-83198-1_1

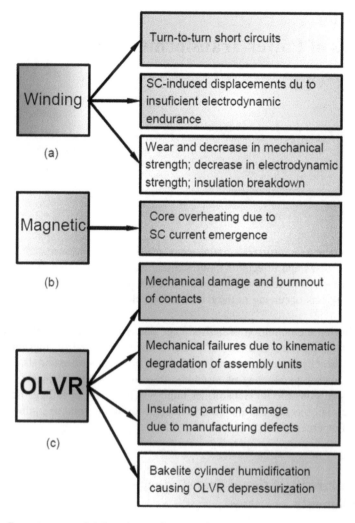

Fig. 1.1 Common types of defects in transformer active parts; **a** winding defects; **b** magnetic conductor defects; **c** OLVR defects

The share of various reasons for transformer failures is as follows:

– 25%—turn-to-turn short circuit;
– 45%—mechanical displacement of windings;
– 30%—other damage.

In some cases (8% of the total), incidents affecting transformers in the specified voltage class and rated power occurred due to reasons unrelated to any winding issues, namely, activation of the gas and differential protection systems of a transformer and gases (mostly acetylene) formation.

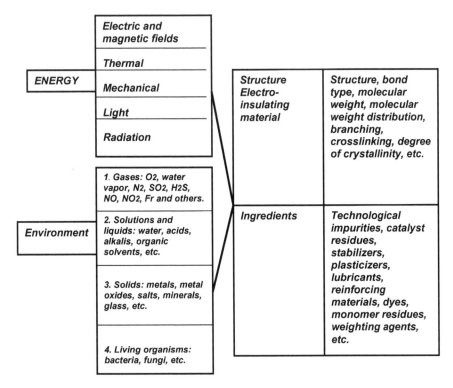

Fig. 1.2 Factors causing the aging of insulation and their dependence on the structure and ingredients of insulation

The next group under study included general-purpose 10/0.4 kV transformers of capacity between 10 and 630 kVA, with natural oil cooling and with a switch without excitation, connected to an alternating current network with a frequency of 50 Hz and intended for power transmission and distribution in moderate and cold climate. Analysis of problem-related situations involving 401 units of 10/0.4 kV 10/(10–630 kVA) power transformers taken out for repair revealed the following: mechanical displacements are responsible for 100 cases, turn-to-turn short circuit explain 74 cases, other defects (oil leakage, switch and contact connection issues, destruction of bushings, burnout of tightening pins) comprise 227 cases. In percentage terms, the result of analysis for 10/0.4 kV transformers is shown in Fig. 1.3 and Table 1.1.

Analysis of the data in Fig. 1.3 and Table 1.1 shows the following:

1. Winding defects account for 43% of the total number of incidents.
2. Mechanical displacements prevail among the winding defects and comprises 25% of the cases; turn-to-turn short circuits in one or both of the windings are responsible for 18% of the incidents in the total number of damaged transformers.

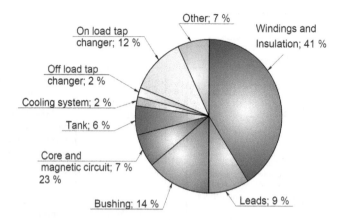

Fig. 1.3 Failure statistics for power transformers [1]

Table 1.1 Failure statistics and main types of damage affecting power transformers of 10/0.4 kV in the distribution grids of one the Russian power systems

Total number of transformers	Number of transformers with mechanical displacement of windings	Number of transformers with turn-to-turn SC	Number of transformers with other damage
200	46	22	132
134	30	10	94
36	13	23	0
31	11	19	1
401	100	74	227

3. Reasons for taking equipment out for repair, other than defective conditions of windings (oil leakage, anzapf switch issues, burnout of contact connections, destruction of bushings, burn out of tightening pins), make up 57%.

The state of events described and exemplified above is typical not only for the energy system of Western Siberia, but also for other electric power systems of the Russian Federation. Thus, according to JSC Samaraenergo, for transformers and autotransformers with a voltage of 110–500 kV, around 30% of the total number of outages has been accompanied by internal SC emergence.

According to conclusions made by a general inspection of 110–500 kV transformers and auto-transformers with a capacity of 63 MVA and above, operated at enterprises of power and intersystem grids in Russia, around 30% of the total amount of technological failures related to disconnection of equipment by protection devices or personnel on an emergency request has been accompanied by internal SC emergence.

The main reasons for such shutdowns are wear and tear, turn-to-turn circuits, breakdowns of taps and winding insulation, insufficient electrodynamics endurance of windings to short circuits, internal insulation breakdowns in high-voltage bushings, and damage to OLVR. One of the main reasons for transformer damage is the insufficient endurance of windings against short circuits, leading to the occurrence of mechanical and electrical defects.

Thus, the analysis of factual material on transformer failures at power system facilities in the Russian Federation shows that one of the main reasons for this state of events are defects in the active parts, namely, the windings, which make up 43% for 10/0.4 kV transformers and over 30% for transformers of the 110 kV voltage class and rated power of 16–40000 kVA. Such statistics are typical for most electric power systems in the Russian Federation.

Detailed accounts of transformer failures in other countries do not appear on a regular basis. According to the Union of German Power Engineers, VDEW, the damage rate for power transformers in 1980–1993 was 0.36% for 110 kV units, 1.54% for 220 kV units, and 2.07% for 380 kV units. Statistics over 13 years for 60 blocks operated in the Netherlands during 700 block-years show a damage rate of about 1.4% per year, with the unit availability factor around 0.995. Calculations show that an increase in the availability factor up to the desired value of 0.998 can be achieved by increasing the transformer stockpile.

The damage statistics of 1970–1986 for large transformers with a voltage of 33–500 kV in Australia revealed a damage rate of around 1% per year. The damage rate in the series of 765 kV transformers operated by American Electric Power (USA) from the beginning of their operation until 1985 amounted to 2.3% per year, while in the series of 345 kV transformers it was 0.7% per year. Based on these data, specialists concluded that it was necessary to develop a new series of 765 kV transformers with an endurance increased to damage, which was accomplished by Westinghouse Electric Corporation. Transformers designed for higher voltage tests replaced the damaged ones. The ANSI/IEEE standard voltage regulations for testing the insulation of HV and LV windings were increased by 14% and 30%, respectively [2].

According to research on the causes and effects of power transformer failures in the Nordic countries, namely, involving Finnish, Swedish, Norwegian and Danish grid companies and conducted from 2000 to 2010 by Vaasa University, Finland, the main reasons for the failures were overvoltage, winding defects and insulation destruction. More than half of the damaged transformers had a service life of 20–40 years [3].

Fig. 1.4 Distribution diagram for transformer defects (Finland)

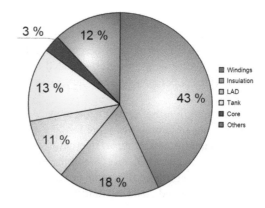

Fig. 1.5 Distribution diagram for transformer defects (Poland)

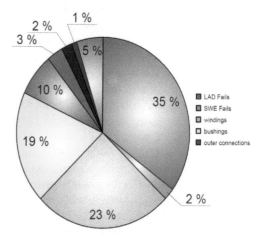

The companies that participated in the study were Finnish (3), Norwegian (6), Danish (1) and Latvian (1). A total of 105 transformers were examined. The most vulnerable components of a transformer proved to be the windings. Almost half of all failures occurred in the windings, OLVR and bushings. Figure 1.4 shows a diagram of damage distribution in percentage terms.

Analysis of the causes of identified internal damage showed that the most frequent defects of 110 kV transformers in Poland were due to local heating of the core and winding, internal short circuits, and simultaneous influence of several types of internal defects [4]. The share of different types of damage is shown in Fig. 1.5 (statistics for 22 years).

It was found that, among all the transformer equipment components, OLVR held the first place in terms of malfunction probability, 35%, while the second place belonged to transformer windings.

Table 1.2 Distribution of damage according to the PN criterion

Damage	PN
Windings	6–48
Insulator	24–48
OLVR	28–52
Core	6
Tank	18
Protection system	22–64
Cooling system	26–48

The results of analysis for the nature and causes of failures in 11 kV/220 V step-down transformers carried out by PESCO, Indonesia, for 5 years (2010–2015) are shown in Table 1.2. The distribution of damage is presented according to the PN* criterion.

* The PN (Priority Number) criterion that determines the degree of urgency for taking action and is described as follows:

$$PN = Severity \times Defect\ Manifestation \times Detection$$

The minimum PN number for any damage is assumed to be 1, with the maximum being 120).

Analysis shows that damage manifests itself most often in the protection system, being 27% of the total amount of damage, and also in the windings, 15%, [5]. The distribution in percentage terms is shown in Fig. 1.6.

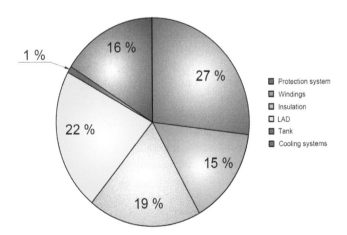

Fig. 1.6 Diagram of transformer defect distribution (Indonesia)

Analysis of statistics on transformer failures in Russia and other countries, as well as research on causes of failure or forced decommissioning of transformers, allows one to conclude that the main indicator of correct and reliable operation of a transformer is the absence of alterations in the geometry of its windings. The latter occurs mainly due to exposure to long-term SC currents, which increases the probability of turn-to-turn short circuits, fires and explosions, with serious consequences. Studies show that around 80% of winding failures in Russia are caused by short circuits. Among the incidents involving internal SC, 24% of the cases were accompanied by transformer ignitions and subsequent fires. The dynamic instability of windings to SC is explained by a weakening sustained with time by the pressing of the core and windings due to mechanical influences in the course of transportation, installation, aging of the insulation, and dynamic forces acting during SC. Windings having some previous displacements show a weakened endurance to subsequent SC currents, which may cause unacceptable damage.

1.2 Technical and Economic Consequences of Failures in Standard Operation of Transformers

The economic damage caused by failures of high-voltage transformers consists of both the cost of failed equipment and the economic damage of consequences inflicted by the accidents. Damage sustained by a single transformer is often the reason for a sudden emergency stop of an entire technological chain involved in an integrated manufacturing process. The result of such an incident is a fire, accompanied by an explosion of high-voltage bushings, an oil spill, environmental pollution and vast economic damage. In this case, neighboring equipment of an electric power system (EPS) may be damaged by scattering of emergency equipment fragments, as well as by high temperatures and other destructive factors. In addition, losses sustained by power systems are increased due to undersupply of electricity, which is an inevitable consequence of accidents. It is also necessary to take into account the environmental damage due to hazardous combustion products released by damaged equipment materials. In this regard, improving the reliability of EPS equipment, including high-voltage power transformers, is one of the most important tasks in ensuring reliable power supply to consumers (Fig. 1.7).

As already noted, the part of equipment that has exhausted its resource in the global electric power industry is increasing, which means the "aging" of electrical equipment, and, primarily, transformers, being the most important and expensive part. In Russia, after a transition between the planned and market economy models, power companies have abandoned a planned system of repair and replacement for electrical equipment and switched on to a "condition-dependent" service system.

Fig. 1.7 Fire in a block transformer of a power plant, the city of Omsk, July 2011. The transformer passed a routine inspection with a complete set of preventive tests and measurements a month before the incident

Recent failures of oil-filled high-voltage equipment are related not only with its moral and physical deterioration. Accidents involving new equipment frequently occur due to a weakening of manufacturing quality control. The probability of such events often exceeds the probability of failure for similar equipment manufactured under strictly controlled conditions, even if it has been in operation for a long time. In this state of events, the task of preventing an operation of newly-made, but low-quality equipment at power industry facilities acquires priority.

The failure of detecting dangerous equipment defects in a timely manner increases the probability of severe accidents, with an increasing amount of repair work and a decreasing service life of equipment. The prevention of serious man-made accidents and catastrophes necessitates the use of even more technologically advanced diagnostic solutions, with a substantiated forecast of the operability of transformers (generally, respective critical structures and equipment). Expert assessments show that up to 80% of defects causing failures of substation equipment and power transmission lines can be timely identified by effective methods and means of monitoring and diagnostics.

Another strategic task is to assess the possibility of continuing the operation of equipment after its pre-assigned service life. The adoption of an optimal decision in this case is based on economic feasibility, which is assessed by comparing the damage resulting from equipment failure with the costs of prevention measures: equipment replacement and diagnostic examination, installation of continuous control (monitoring) systems, recovery (current or capital) repairs based on the results of a diagnostic examination.

In this regard, the immediate task of the global electric power industry is the development of effective diagnostic systems, methods and diagnostic tools that make it possible to confirm the operability of equipment (defect-free condition), or detect damage at an early stage of development, and, ultimately, estimate the residual resource and take measures to increase the latter.

Below we examine two main groups of reasons for transformer failures in case no timely prevention measures have been taken: (1) insulation degradation, (2) integrity violation of transformer windings.

1.3   Insulation Aging (Degradation)

During operation, a prolonged exposure to operating voltage and short-term exposure to overvoltage, combined with other factors (elevated temperature, mechanical stress, environmental conditions, etc.) leads to a decrease in the dielectric strength of insulation and its failure due to aging and a subsequent breakdown. Aging is the development of an involved combination of physical, chemical, electrical and other processes in the insulation [6]. Each of them, at a sufficiently high intensity, can be registered and studied due to certain external manifestations being characteristic only for a specific process. This is the basis of present-day methods to study the aging of electrical insulation and monitoring of its current state.

As regards the combined insulation of high-voltage transformers (paper-oil, oil-barrier, paper-film-oil) characterized by a high intensity of aging processes, some methods of condition monitoring have been developed and are effectively used at present. They are based on registering partial discharges (PD), controlling the intensity of gas evolution and alteration of chemical compositions, as well as measuring the dielectric loss tangent (tan δ), insulation resistance and absorption currents, etc., [7, 8].

Taking into account a large PD contribution to the damage and degradation of transformer insulation, as well as a complexity of the related characteristics and their nature, below we examine all of them in sufficient detail. Other degradation mechanisms for transformer oil and cellulose (chemical, thermomechanical, etc.) are considered in the sections devoted to a brief description of control methods on a basis of their research and registration.

PDs in gas pores, gas wedges and interlayers provide the main cause of degradation and failure of transformer insulation. In choosing permissible electric field strengths in such and similar insulation (capacitors, bushings, cables with monolithic insulation), PD characteristics are decisive. Material discontinuities, hereinafter called "inclusions", are formed in the course of manufacturing electrical insulating materials and insulating structures on a basis of the former, which makes such inclusions inevitable in practice. Once an electric field is applied to insulation, ionization processes are initiated in the inclusions, being limited by the insulation volume, and thereby called *"partial discharges"* (PD). A primary PD occurrence in the inclusions is caused by two reasons: (1) a lower electrical strength of the medium filling the inclusion, (2) increased electric field strength in the inclusion caused by its redistribution due to a difference in the dielectric constant of the base material (higher) and the material of the inclusion (lower). Electron-avalanche processes in inclusions begin with auto-electron emission from an electrode (wire) in case the electrode adjacent to the cavity is of negative polarity, or, in the case of holes being emitted once the electrode polarity is reversed, as well as in the case of closed pores. PD in cavities ranging in size from a few microns to fractions of a millimeter is basically a Townsend discharge. However, the duration of a single PD (hundreds of picoseconds), being a very short one, given a pore size (ranging from tens to hundreds of microns), indicates that secondary emission is due to the action of photons rather than positive ions. The threshold energy and quantum yield of photoelectron emission from the surface of an organic dielectric are not very different from those in metals: in dielectrics, the threshold energy is slightly higher, and the quantum yield is lower. For this reason, all other things being equal, the breakdown voltage of a gas microcavity in a dielectric is somewhat higher than the microgap between the metal electrodes. When the size of a cavity is tens of microns and the gas pressure in it is close to atmospheric, the breakdown voltage is near the minimum of the Paschen curve, i.e., is 250–300 V.

An electric discharge in a gas cavity of larger size is a streamer discharge with its characteristic features: a contracted channel, a high rate of development (elongation) of the channel, and a high plasma temperature [9].

The destruction of walls in micron and submicron cavities is caused not by the development of a partial discharge, but rather by the bombardment of cavity walls by emitted electrons. Thus, the size of a cavity and the gas pressure inside of it determine the mechanism of processes developing under the field action, as well as the intensity of material destruction. The prevalence of one or several factors of insulation destruction over the others is determined by many variables: field strength, dielectric properties, gas cavity parameters [10]. In particular, an increase in the PD intensity in relatively large cavities with an increase in the field strength leads to the fact that the process of point erosion in the cavity walls begins to prevail over the oxidation process. An increase in the rated energy release and the rated power of the PD is one of the reasons for a low long-term electric endurance of dielectrics with the cavities elongated along the force lines of the electric field.

If there are points of significant increase in the electric field (micropoints) on the electrode (wire), PDs can appear even in the absence of macroinhomogeneities (inclusions) in the insulation at the initial state. (For transformer insulation, this case if unlikely). Then, charge carrier emission, ionization, panderomotive forces, and thermodynamic processes may lead to cavity emergence, accommodating PD occurrence [11]. PDs cause erosion of pore walls and gradual destruction of insulation with its subsequent through breakdown. For a significant thickness of solid insulation, predominantly polymeric, PDs can degenerate into electrically conductive tree-like or bushy shoots, *dendrites* [12].

The entire sequence of processes, from the formation of inhomogeneities (when they are absent at the initial state) to the breakdown of insulation, is shown in Fig. 1.8.

During the combustion of a discharge in the channels of a dendrite, the electrical conductivity is provided by gas-discharge plasma filling the channels, and during the quenching of the discharge, by carbon deposits on the walls of the channels, as a result of the thermochemical decomposition of organic insulation. Under certain conditions, the channel cavities are filled with liquid. Such dendrites, observed mainly in monolithic polymer insulation, are called *water dendrites.* They are unlikely to occur in transformer insulation and, are therefore not considered here, or in Subsection 2.7. Information in their respect can be found in [13].

A certain variety of dendrites is regarded as *sliding* **or** *creeping discharges* that develop along the surface of solid insulation surrounded by a liquid or gaseous medium [14–16].

Despite the fact that they develop in an environment, their plasma channels cause erosion and carburization of the surface layers in solid insulation, namely, the formation of conductive tracks. The tracks slowly penetrate the insulation and at some stage lead to the growth of a dendrite through the insulation. Dendrites and creeping discharges are shown in Fig. 1.9.

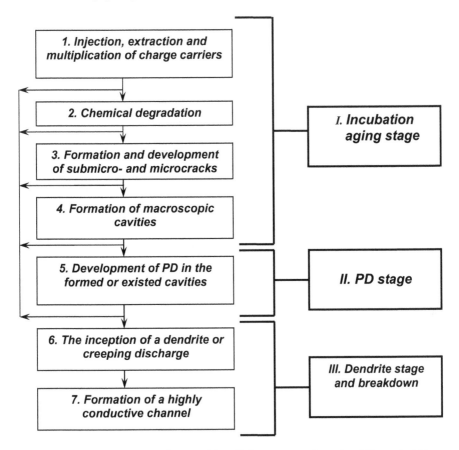

Fig. 1.8 Sequence and mutual influence of degradation processes for an insulating material in an electric field

Establishing the emergence of PDs and measuring their characteristics makes it possible to assess the quality of insulation manufacturing and to identify local defects both in the insulation and on the current-carrying parts of a structure. In registering PDs during the operation of a high-voltage device (for example, a transformer), it is possible to control the rate of natural or accelerated aging of this device. PD characteristics correlate fairly well with the size and number of defects, i.e., they allow making a reasonable conclusion about the degree of defectiveness of an insulating structure at different operation stages for a high-voltage apparatus. The study of PD characteristics is a task of paramount importance for all high-voltage devices, and, primarily, for those in which multilayer (hence, non-uniform) insulation is used [17].

When considering the mechanism of PD initiation, an equivalent circuit with a total capacitance C_E is used to replace a dielectric, Fig. 1.10.

Fig. 1.9 Photo of a dendrite in a translucent polymer dielectric

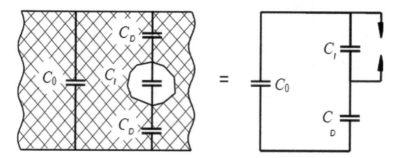

Fig. 1.10 Equivalent circuit for a solid dielectric: C_0 is the capacity of non-defective insulation; C_I is the capacity of an air inclusion; C_D is the dielectric capacitance in series with the inclusion; U_I is the breakdown voltage of the air inclusion

$$C_E = C_0 + \frac{C_I \cdot C_D}{C_I + C_D}. \tag{1.1}$$

PDs occur when the switch-on voltage reaches the breakdown value U_I, in this case being the discharge ignition voltage at the switch-on. The electric field strength in the inclusion E_I is related to the strength in the rest of the dielectric as follows:

$$E_I = E_D \cdot \frac{\varepsilon_D}{\varepsilon_I}, \tag{1.2}$$

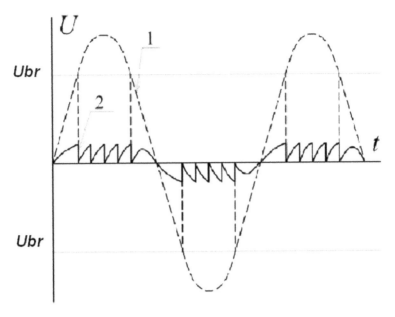

Fig. 1.11 Diagrams of voltage on an air inclusion in a solid dielectric: 1 is the voltage on the sample; 2 is the switch-on voltage; U_{br} is the voltage on the sample at which an air inclusion breakdown takes place

where: E_D is the electric field strength in the dielectric, ε_D is the relative dielectric constant of the dielectric, and ε_I the relative dielectric constant of the inclusion.

Based on (1.2), the electric field strength in a gas inclusion (or any other inclusion with $\varepsilon_I < \varepsilon_D$) is always higher than it is in the rest of the dielectric. The voltage diagrams at the switch-on when alternating voltage is applied are shown in Fig. 1.11.

For an inclusion size of tens of micrometers and a pressure close to atmospheric, the breakdown voltage is near the minimum of the Paschen curve, while exhibiting small changes with a change in the inclusion size, and being equal to 250–300 V.

PDs are especially hazardous at alternating or impulse voltages.

PD can be classified by qCR value in the following way.

1. When a certain voltage threshold is exceeded, PDs appear in insulation with an intensity of $q_{PD} = 10^{-12}$–10^{-11} C. Such PDs do not cause rapid destruction of insulation, and, in many cases, are acceptable for long periods of equipment operation. Such PDs are referred to as ***initial***.

2. Any further increase in voltage, or an increase in the size of inclusions in the course of long-term operation of the insulation, leads to a sharp increase in the PD intensity, primarily with q_{PD} increasing up to 10^{-10}–10^{-8} C. The occurrence of such PDs drastically shortens the life of insulation, and thus they are called ***critical***. Dendrites can be regarded as a final stage in the development of critical PDs.

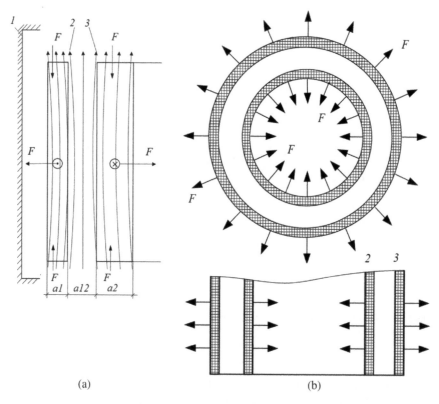

Fig. 1.12 Axial forces in longitudinal section **a** radial forces in a cross section **b** 1 is the magnetic circuit of a transformer; 2 is the inner winding; 3 is the outer winding

Under the influence of critical PDs in oil channels, the process of destruction is intensified, initially in oil-barrier insulation and then in coil insulation, at the locations of inclusions with the emergence of damaged areas. This is intensified by the heating of the oil caused by the closure of the magnetic conductor plates. Destruction processes take place in both the oil-barrier insulation and the high- and low-voltage insulation windings. In the area of the coil insulation sections that have undergone destruction, there occurs a strong heating of adjacent insulation sections and transformer oil; in the heated oil there emerge bubbles supporting PD occurrence. Sliding discharges begin to form along the dielectric surfaces. Sliding discharges on the surface of barriers and coil insulation, as well as PDs in oil interlayers, and corona-shaped PDs in a bare edge area of the winding wires cause significant heating of the oil. This heating, combined with heating due to the plates closure (*steel fire*), increases the destructive effect on all the elements of insulating structures. The total heating leads to an increase in the intensity of all electrophysical and thermochemical processes in the insulation, which reinforce each other and accelerate the rate of destruction sustained by the insulation (accelerated aging), resulting eventually in a complete electrical breakdown and destruction of the entire electrical insulation system of a transformer.

Fig. 1.13 Radial displacement of winding turns

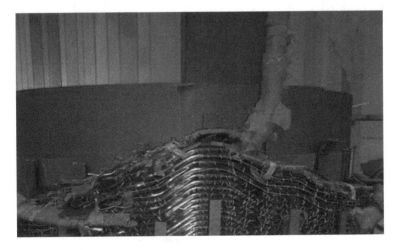

Fig. 1.14 Complex displacement of winding turns

1.4 Integrity Violation of Transformer Windings

Mechanical defects of windings come in two main types: axial (longitudinal) and radial (transverse) displacements. Diagrams exemplifying the radial effect of ponderomotive forces are shown in Fig. 1.12.

Ponderomotive forces acting on a winding cause a radial displacement of the winding turns, Fig. 1.13, with a subsequent complex displacement of the turns, Fig. 1.14.

Diagrams exemplifying the tangential effect of ponderomotive forces are shown in Fig. 1.15.

The tangential component of ponderomotive forces acting on a winding causes an axial displacement of winding turns, Fig. 1.16, with a subsequent defect of "lodging of conductors", Fig. 1.17.

A defect of the "short-circuited turn" type occurs when the turn insulation is damaged by SC currents or switch over-voltages. The appearance of a "short-circuited turn" defect is shown in Fig. 1.18.

One of the main causes of damage to transformer windings (dynamic instability deformation) is due to SC currents in power grids. Currently, such damage to transformers has a prominent position. Calculations show that approximately 1.7% of 220–500 kV autotransformers may be exposed once a year to dangerous SC current effects, which are especially hazardous for autotransformers with reduced electrodynamic stability. For example, at the substations of RAO UES such a "risk group" is

Fig. 1.15 Impact of tangential forces on a winding

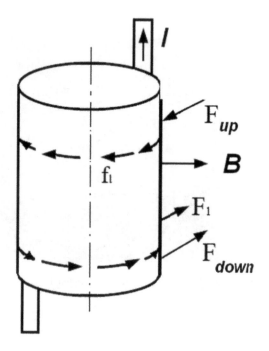

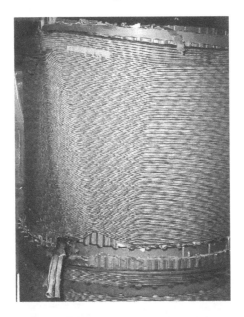

Fig. 1.16 Axial winding displacement

estimated at 25% of the total number of autotransformers in the 330–750 kV voltage classes. Recently, significant long-term voltage rises in power grids (abnormal operating modes in energy systems) have become a factor of danger for power transformer operation. The problem of dynamic instability of winding design has also become more serious due to a steady growth of SC intensity in power systems.

Fig. 1.17 Defect of the
winding type "lodging of
conductors"

Fig. 1.18 "Short-circuited
turn" type of winding defects

References

1. Fuhr J, Aschwanden T (2017) Identification and localization of PD-sources in power transformers and power generators. IEEE Trans Dielectr Electr Insul 24(1):17–30
2. https://www.transform.ru/sst/usege/ss/alexeev/oee.htm
3. Taikina M (2010) Survey of transformer failure causes and consequences. Bachelor's thesis in Technology for the degree of Bachelor of Science in Technology submitted for inspection in Vaasa, 7th of September, 2010
4. Domzalski T (1997) Reconstruction of transformers on site. Trans Distrib World 2:43–38
5. Jan ST, Afzal R, Khan AZ (2015) Transformer failures, causes and impact. International conference data mining, civil and mechanical engineering (ICDMCME'2015) Feb. 1–2, 2015 Bali (Indonesia)
6. Ushakov VYa (1993) Aging of insulation and methods of condition monitoring. TPU Publishing House, Tomsk, 60 p (in Russian)
7. Elektrische Isoiertechnik KM (1988) 1 Ausgabe. VEB Verlag Technik, Berlin, 360 S
8. Svi PM (1980) Insulation monitoring of high voltage equipment, Energiya, Moscow, 112 p (in Russian)
9. Dzhuvarly ChM, Vechhaizer GV, Leonov PV (1983) Electric discharge in gas inclusions in high-voltage insulation, Elm, Baku, 192 p (in Russian)
10. Ushakov VYa (1988) Electrical aging and service life of monolithic polymer insulation. Energoatomizdat, Moscow, 152 p (in Russian)
11. Zaengl W, Kuffel E, Kuffel J (200) High voltage engineering. Fundamentals. Butterworth Heinemann, 539 p
12. Ushakov VYa (2004) Insulation of high-voltage equipment. Springer Verlag, Berlin, Heidelberg, 421 p
13. Moreau E, Mayoux C, Laurent C (1993) the structural characteristics of water trees in power cables and laboratory specimens. IEEE Trans Electr Insul 28(1):54–64. https://doi.org/10.1109/14.192240
14. Ushakov VYa, Klimkin VF, Korobeynikov SV (2007) Impulse breakdown of liquids. Ed. by Usakov VYa, Springer Verlag, Berlin, Heidelberg, 397 p
15. Murdiya F, Hanaoka R, Akiyama H, Miyagi K (2012) Creeping discharge developing on vegetable-based oil pressboard interface under AC voltage. IEEE Trans Dielectr Electr Insul 21(5):2102–2110
16. Sadaoui F, Beroual A (2014) DC Creeping discharges over insulating surfaces in different gases and mixtures. IEEE Trans Dielectr Electr Insul 21(5):2088–2094
17. Kuchinsky GS (1979) Partial discharges in high-voltage structures, Energy, Leningrad, 1979–224 p (in Russian)

Chapter 2
Standardized Technologies of Condition Monitoring for High Voltage

For monitoring the condition of transformer active parts, there are two main requirements that stand out: detection of defects at an early stage and diagnostics with voltage in operation and transformer bundling intact ("ON-LINE" mode). The monitoring of defects at an early stage makes it possible to largely extend the resource and service life of a transformer and to avoid unforeseen accidents that may cause severe technical catastrophes (fires, explosions) and economic losses (shortage of electricity, power outages, financial losses in fines). To meet such demands, advanced power systems show a growing tendency of moving away from regulated periodic technical examinations of transformers to their monitoring, with disconnection made to address the condition of transformer elements, in order to carry out a complex diagnostic test using methods that cannot be applied to a transformer in operation.

Technologies (methods) for monitoring the condition of a transformer can be divided systematically into the following two largely different groups: (1) non-electrical and (2) electrical.

Non-electrical methods include:

- physical and chemical control,
- vibration control,
- acoustic control (used mainly for PD registration, it will be discussed in Sect. 2.2.1).

Electrical methods include:

- PD registration,
- measurement of idling losses,
- measurement of the transformation ratio,
- monitoring the insulation resistance of transformer windings,
- measurement of the ohmic resistance of windings,
- control of changes in SC resistance (inductance),
- registration of responses in pulse probing of transformer windings.

V. Ya. Ushakov et al., *Transformer Condition Control*, Power Systems,
https://doi.org/10.1007/978-3-030-83198-1_2

2.1 Non-electrical Control Methods

2.1.1 Physicochemical Control Methods

Physicochemical processes developing in insulation during the operation of oil-filled transformers can be examined by indicators used to estimate transformer conditions [1].

The acid number is the amount of caustic potassium (ACP), measured in milligrams, that is required to neutralize free acids in 1 g of oil. An increase in the acid number indicates the oxidation of the oil, which may cause corrosion of structure elements, emergence of soaps with metal ions, and development of colloidal-dispersed processes leading to a decrease in the electrical strength of the oil. Acids, due to their polarity, can also help increase the water absorption of paper insulation.

The content of water-soluble acids and alkalis that can form during oil manufacturing or as a result of oil oxidation in the course of operation. These ingredients are quite corrosive and contribute to the development of corrosion and the aging of paper insulation.

The moisture content, as an indicator of solid insulation and oil conditions, which is periodically monitored during the operation of transformers. Atmospheric moisture penetrates into the oil or emerges in it due to the following reasons: (1) absence or malfunction of dryers in transformers with free breathing, (2) suction of moist air or rain water into the oil in transformers with a forced cooling system when the latter is leaking, (3) suction of humid air through other leaks, (4) formation of moisture as a result of aging processes in insulation itself.

The gas content of the oil is monitored in operating transformers with a film protection of the oil from oxidation in order to estimate transformer tightness. An increase in the gas content (including the air) promotes more intensive oil oxidation and a decrease in the dielectric strength of insulation in the active part of a transformer.

The above physical and chemical indicators have been used for many years in traditional methods for diagnostics of power transformer conditions, elaborated in the vast literature available to English-speaking readers [2–9]. Therefore, such indicators are not considered in this monograph.

2.1.1.1 Chromatographic Analysis of Dissolved Gases

Chromatographic analysis of dissolved gases is the most common type of monitoring, used in almost every country. All major electrical companies and transformer manufacturing enterprises widely use the analysis of gases dissolved in oil, combined with diverse systems for evaluating and identifying the type of transformer defects.

This method (abbreviated as CADG) is attractive by its high sensitivity to electrical discharges in insulation (PD, dendrites, creeping discharges) and to local overheating, responsible for decomposition of paper insulation and oil. Special attention is paid to the content of furan derivatives, which may be indicative to the destruction of paper insulation. Thermolysis, oxidation and hydrolysis of insulation, causing partial destruction of cellulose macromolecules, lead to the emergence of furan series components released into transformer oil.

Regulatory documentation used since the 1980–1990s in all countries with a developed electric power industry provides for the CADG method to estimate the paper insulation condition for power transformers in operation. Since that time, a fairly large experience in using the method has been acquired in relation to power transformers with a voltage of 110–750 kV. This experience makes it possible to select a number of indicators that have a relatively high diagnostic value, and to determine the type and nature of defects identified for making decisions on further operation of a transformer.

The CADG method can be used to identify two groups of defects in power transformers:

– overheating of current-carrying connections and frame structure elements;
– electrical discharges in oil.

The concentration of the following seven gases is then determined: hydrogen (H_2), methane (CH_4), acetylene (C_2H_2), ethylene (C_2H_4), ethane (C_2H_6), carbon monoxide (CO) and carbon dioxide (CO_2).

Defects in power transformers are identified using a division of gases into basic (key) and accompanying (characteristic) gases.

In the case of overheating in current-carrying connections and transformer frame elements, the basic gases are C_2H_4 (with oil-paper insulation and oil being heated above 500 °C) and C_2H_2 (during an arc discharge). The accompanying gases in both cases are H_2, CH_4 and C_2H_6.

In the case of partial discharges in oil, the main gas is H_2, while the accompanying gases with a low content are CH_4 and C_2H_2.

In spark and arc discharges, the basic gases are H_2 or C_2H_2, while the accompanying gases with any content are CH_4 and C_2H_4.

In the case of solid insulation overheating, the main gas is CO_2. It should also be noted that a concomitant indicator of destruction for cellulose insulation in a transformer is an increase in the content of oxide and carbon dioxide dissolved in transformer oil. At the same time, in accordance with the recommendations of The International Council on Large Electric Systems (CIGRE), the presence of a total CO and CO_2 concentration of more than 1% may indicate the degradation of cellulose insulation [10].

The basic gas is determined by relative concentrations of hydrogen and hydrocarbons, taking into account the corresponding boundary concentrations according to the formula

$$a_i = A_i / A_{rpi} \tag{2.1}$$

where a_i is the relative concentration of a selected gas; A_i is the measured value of the selected gas concentration, % vol; A_{rpi} is the boundary concentration of the selected gas in volume percent (% vol).

According to calculated relative concentrations, the maximum value of a_i corresponds to the basic gas, except for CO_2, being a basic gas if $A_{CO2} > \%$ (vol). In this case,

$a_i > 1$, characteristic gas with a high content;
$0.1 < a_i < 1$, characteristic gas with a low content;
$a_i < 0.1$, uncharacteristic gas.

It should be pointed out that in analyzing the composition and concentration of gases dissolved in oil, as well as in diagnosing the operational condition of power transformers, it is necessary to take into account the factors that cause transformers to experience alterations.

The operational factors causing an increase in the concentration of gases dissolved in transformer oil include: residual gas concentrations from a previous defect eliminated during repair (the oil has not been degassed); increasing the load of a transformer; topping up with used oil containing dissolved gases; performing welding work on the tank; overflow of gases from the expander tank of an OLVR contactor to the transformer tank. The operational factors causing a decrease in the concentration of gases dissolved in transformer oil include: a decrease in the load of a transformer; oil degassing; topping up with degassed oil; replacement of silica gel.

To diagnose defects developing in power transformers, the following main criteria are used:

- criterion of boundary concentrations;
- criterion of gas growth rate;
- criterion of pair ratios for characteristic gases.

According to regulatory documentation, the values of parameters exceeding some predetermined limits (limit values) should be regarded as indicators of defects that may lead to equipment failure. In this regard, the CADG method is different from the above ideology, since normative boundary gas concentrations are those whose occurrence indicates a mere chance for defects to develop in a transformer [11]. Such transformers are taken under control with increased oil samplings and CADG procedures.

The criterion of boundary concentrations permits selecting transformers with possible defects in development from the total number of units in a transformer stock, while the hazard degree for a defect to develop is determined by the relative growth rate of gas (gases). The amount of accumulated experience shows that once the relative growth rate of gas (gases) exceeds 10% a month, this indicates the presence of a rapidly developing transformer defect.

The nature of defects in development, according to CADG results, is determined by a criterion of concentration ratios for various gas pairs. It is customary to distinguish between thermal and electrical defects. Defects of thermal nature include the occurrence of short-circuited circuits, increased heating of insulation and contacts of on-load voltage regulators (OLVR) and those of control (switching) without excitation (SWE), taps, pins, and other frame and tank metal structures of a transformer. Defects of electrical nature include discharges in oil of various intensities. Naturally, the development of transformer failures may be of mixed character. Analysis of methods available for estimating the nature of defects in development (thermal or electrical ones), based on the results of CADG, shows significant differences in both types and gas pair ratios used. These ratios have been established in a number of techniques: Dörnenburg's method, Müller's method, CEGB/Rogers Ratios, Schliesinger's method, normogram method, Duval's method, IEC 60599.

Methods of interpreting CADG results for transformers used by companies in different countries are noticeably different. Thus, ABB has adopted the Dörnenburg's method using three basic gas ratios, according to IEC, with a set of four types of defects (thermal, partial discharges of low and high energy, powerful discharges) given a simple graphic interpretation. Limits of gas concentration have been recommended for transformers without defects, regardless of the time of their operation.

Spain has adopted the ASINEL system applying seven digital codes assigned to six gas ratios and hydrogen concentrations, as well as to a sum of the C1 and C2 hydrocarbons. Three of the codes correspond to IEC 60599 Codes and have the same meanings; one of the gas ratios corresponds to the Rogers Code, and another to CO_2/CO, adopted for estimating the degradation of paper insulation. The ratio C_2H_2/H_2 is used to estimate the location of a failure: once it is higher than 2, there is an OLVR defect, and once lower, the defect is located in the main tank of a transformer.

In Canada, the diagnosis is made using the Duval triangle based on relative concentrations of the H_2, C_2H_2, C_2H_4, C_2H_6 gases for six main types of defects. The criteria for gas concentrations under normal, acceptable and hazardous conditions are significantly varied for different gases and, in particular, for different values of voltage and service life of transformers. To obtain a confident diagnosis, CADG data are entered into an expert system for estimating the condition of a transformer.

A detailed analysis of CADG efficiency for transformers has been made by the Canadian IREQ Institute. It is based on 25,000 samplings taken by the energy company Hydro Quebec [12].

The system by the LABELEC Laboratory (Portugal) is based on the IEC 599 gas ratios, as well as on the CO_2/CO and C_2H_2/H_2 ratios; however, it provides for a first-priority assessment of absolute values for gas concentrations. If the volume of at least one of the gases goes beyond the limit value, then five gas ratios are calculated. Additionally, the Schliesinger method is used to confirm the presence of more than one type of simultaneous defects.

RWE Energie (Germany) has adopted a system using five gas ratios and four limit concentrations. Depending on the value of a ratio, the severity of a defect is identified and compared with the concentration of gases. Significant basic gas ratios and concentrations are regarded as indicators of a significant stage of insulation degradation. The system distinguishes between five types of defects. A similar system (by KEMA Transformer) is adopted in the Netherlands.

Belgium has adopted a system based on the Laborelec Code Table. The codes represent concentrations of the H2, C2H2, CO gases and the sums of C1 and C2 hydrocarbons. In addition, the CH4/H2 ratio is taken into account. The system includes nine ways of identifying defects being distinct in severity. Depending on concentrations and gas ratios, 27 different types of diagnosis are provided, indicated by codes. The normal condition of a transformer corresponds to a hydrogen concentration of up to 200×10^{-6} rel. units, with the sum of C1 and C2 hydrocarbons being $300 \cdot 10^{-6}$ rel. units, and CO being 800×10^{-6} rel. units.

France has adopted the system by the LCIE Laboratory, based on comparing gas concentrations with a standard, taking into account the voltage, type of design (presence or absence of OLVR), terms and conditions of transformer operation. Once concentrations go beyond the standard values, gas ratios are analyzed to identify the type of defects. The rate of increase in gas concentrations is taken into account as well. The LCIE Laboratory also uses C_3 hydrocarbons to interpret the results of CADG.

The National Grid energy company (Great Britain) traditionally uses the Rogers Ratios method. It is believed that the C2H6/CH4 gas ratio, excluded from the IEC 60599 system, is efficient at identifying an increase in temperature as compared to an operating temperature. The method also takes into account an increase in gas emission over time, and a comparison of gas concentrations with the admissible one. The method continues to improve.

Siemens Trafo Union (Germany) uses concentrations of characteristic gases and their ratios as criteria. The concentration of acetylene, hydrogen and that of a sum of C_1 and C_2 hydrocarbons is analyzed for propylene, whereas the sum of CO and CO_2 is used for propane.

In Poland, gas concentrations and ratios of characteristic gases are also used as criteria taking into account increase rates for gas content in oil.

It is noteworthy that admissible concentrations of gases in oil, determined from practical experience, are assumed to be different for block and grid transformers, being 10^{-6} units, see Table 2.1.

Energopomiar, implementing CADG, annually conducts over 650 oil samplings from large transformers. CADG is also widely implemented by Elta Manufacturing, being the main transformer enterprise in the country [13].

According to the results of CADG [14, 15], it has been found that the IEC 60599 method, recommended for use in [16], offers the highest diagnostic capacity of identifying the nature of defects in development.

The practice of Russian power systems relies on gas chromatography as a basic method for estimating the technical condition of transformer equipment. Criteria are distinguished by the assignment of equipment to frequent control (limit concentration criterion), the type of defects in development (concentration ratio criterion), and the danger of defects (rise rate criterion). To identify thermal defects at early stages of their development, oil analysis is carried out to determine the presence (additional to the seven basic gases) of C_3 and C_4 hydrocarbons. For this analysis, the gas chromatographic complex Tsvet 500-TM is recommended, as well as the complex manufactured by OOO NPF "Elektra".

Table 2.1 Admissible concentrations of gases in oil for block and grid transformers

Gas	Block transformers	Grid transformers
Hydrogen H_2	260	500
Methane CH_4	250	200
Ethane C_2H_6	160	170
Ethylene C_2H_4	250	260
Acetylene C_2H_2	20	70
Propane C_3H_8	40	30
Propylene C_3H_6	40	40
Carbon monoxide CO	280	260
Carbon dioxide CO_2	3500[a]	4000

[a]Higher concentrations of CO_2 are allowed only when the CO/CO2 ratio does not exceed 0.3

The experience of running CADG on transformers operated by various enterprises shows that it is advisory to take into account different factors when interpreting the results of analysis. Taking into account changes in gas concentration when comparing the results of two samplings conducted at different operation periods makes it possible to identify defects in development and estimate their danger. It should be noted that comparison of absolute gas release values for different transformers necessitates a reduction to one and the same oil volume. In particular, this technique is applied to study the on-load dependence of gas release.

Positive results are ensured by analyzing thermodynamic processes in a transformer prior to its damage: at the location of defect development there is a zone where the temperature is sufficient for releasing the primary decomposition products of complex oil molecules. These products penetrate into the cooling zone, where equilibrium is achieved. The relative concentration of constituent gases can be calculated as a temperature function and then applied to estimate the change.

In many systems, the presence or absence of OLVR is duly taken into account; the sampling of oil is made on a regular basis; the rate of increase in gas concentration is accounted for, and the use of CADG at the initial period of transformer operation is duly distinguished. In determining the limit values of gas concentration, one takes into account the voltage class, the load intensity, and the possibility of gases penetrating from the OLVR tank into the main transformer tank.

To interpret the results of analysis, other criteria are also used, developed using statistical and practical operating data. For instance, such criteria are given by an increase in gas concentration following 100 h of transformer operation (according to Russian and Ukrainian research), and by a difference between grid and block transformers.

The lack of a unified methodology for interpreting the results of CADG in power transformers makes it difficult to compare the condition of transformers operated and controlled by different enterprises. It is difficult to coordinate most of the criteria for

Table 2.2 Normal concentrations of basic gases according to the innovative method CIGRE WG 15.01

Basic gas, 10^{-6} rel. units	Excess concentration	Possible defect
C_2H_2	20	Intense discharges
H_2	100	Partial discharges
C_XH_Y sum		
C_1, C_2, C_3 gases	1,000	Thermal defects
C_1 and C_2 gases	500	Thermal defects
CO_2 and CO gases	10,000	Cellulose degradation

assessment of transformer conditions and put to use the experience of other power industry participants.

Based on efficiency analysis for monitoring transformer conditions by different methods using CADG, the CIGRE Working Group 15.01 has come up with an innovative method for interpreting CADG results, *CIGRE WG 15.01*. For large network and block transformers, the method normalizes the concentration of basic (key) gases, see Table 2.2.

The new method provides the following key gas ratios:

Ratio No. 1—C2H2/C2H6 (acetylene/ethane). Ratio No. 1 is considered to be crucial in determining the presence of electrical discharges; a value larger than a unity indicates a defect.

Ratio No. 2—H2/CH4 (hydrogen/methane). The presence of partial discharges is determined in relation to No. 2. Usually, the value is more than ten. (IEC 60599 uses a methane/hydrogen ratio).

Ratio No. 3—C2H4/C2H6 (ethylene/ethane). Ratio No. 3 is the ratio of unsaturated hydrocarbons to saturated hydrocarbons and reveals thermal effects. Usually, the value should be larger than a unity. Unsaturated hydrocarbons are formed mainly in overheated oil.

Ratio No. 4—CO2/CO (carbon dioxide/monoxide). Ratio No. 4 determines the degree of cellulose degradation. If the value is more than ten, then cellulose overheating takes place. If the value is less than three, then it is usually regarded as evidence of cellulose degradation under the influence of electrical defects. A furan analysis according to IEC 61198 is recommended to confirm the diagnosis.

Ratio No. 5—C2H1/H2 (acetylene/hydrogen). Ratio No. 5 is used to determine the penetration of gases into the common tank from the OLVR tank. In this case, the ratio usually equals two or more units, and the C_2H_2 concentration is no less than $30 \cdot 10^{-6}$ rel. units. Since hydrogen is less soluble in transformer oil as compared to acetylene, the latter diffuses from the OLVR tank more rapidly; only a small amount of hydrogen diffuses into the main tank. As a result, the amount of acetylene in the transformer oil exceeds that of hydrogen.

CADG result interpretation procedure

- ratios and concentrations of key gases dissolved in oil are identified;
- if all concentrations are below normal ones, then the result is indicated by K1. If at least one of the concentrations has exceeded the normal range values, then code K2 is assigned;
- if all key ratios are below their specified limits (inside the limits, in the case of CO and CO_2), the result is indicated by code E1;
- if any of the ratios exceeds the specified limits (reaches outside the limits, in the case of CO and CO_2), the result is indicated by code B2.

Combined results:

(a) K1 and E1: no measures are taken; the transformer most likely has no defects;
(b) K2 and K1: the transformer is likely to have a defect; further analysis is required;
(c) K1 and K2: defects in development are possible; further analysis is required;
(d) K2 and B2: more than one type of defect is likely; further analysis is required.

For the reliability of a diagnosis, sampling is repeated to calculate the above combinations of codes, with the exception of the first one. Further analysis and other examinations are carried out to find the nature of defects in occurrence. The obtained data are compared with statistical distributions of the probability of a transformer defect—as dependent on the content of H_2, CH_4, C_2H_2, C_2H_4, C_2H_6, N_2, O_2 gases dissolved in oil—available in the databases of various organizations for different voltage classes, workload and service life of transformers. These diagrams show, for different voltage classes and workload levels of a transformer, that the hydrogen concentration does not affect the distribution pattern. For the distribution of a sum of hydrocarbons (C_XH_Y), such dependence does actually exist. The distribution of CO_2 is influenced by voltage classes; CO_2 concentrations are much higher for block transformers, normally operated at base load, than they are for grid transformers.

Research is in progress to determine the "normal" values for various classes of transformers. The innovative method CIGRE WG 15.01 is characterized by a step-by-step assessment of transformer conditions: first, one uses simple methods to determine a possible presence of defects and a need for further analysis of the transformer, and then the nature and severity of defects are identified by a more detailed analysis of CADG data. The new method not only complies with the IEC 60599 recommendations, but also develops them in a way suitable for practice.

According to the Transformers Gas Analyst™ method by Delta-X Research, division of Hydra-Centaurus Technologies Inc (Canada), results of analysis and trends of changes are displayed, and control data are documented. The database also contains monitoring rules and instructions, and thereby makes it possible to reflect the experience of equipment operation and operating conditions.

Analysis of gas from gas relays is regarded by some foreign enterprises as a particularly effective direction for diagnosing the condition of a transformer. Sampling is made when a gas relay is triggered by a signal; then diagnostic methods are used to determine the amount, location and nature of the defect.

Taken together, the results of CADG are indications for extraordinary measurements of winding insulation resistance, dielectric loss angle tangent, winding resistance to direct current, idling losses, thermal imaging control of transformer tank and cooling system surfaces, as well as for execution of CADG in contactor tank oil.

Based on the amount of measurements and their results, a number of questions can be answered:

- the amount and duration of overloading the transformer;
- the temperature in the upper layers of oil and air;
- the possibility of a decrease in the oil level relative to the top of the radiator;
- the operability of the transformer fans and radiators
- the orientation of the transformer in relation to the sunrays;
- the fact of the transformer being closed on three sides without blowing.

Based on the results, one of the following decisions, or a number of them, must be taken:

- retain the transformer in operation under enhanced CADG control;
- verify the operation of the oil pump engines;
- verify the presence of gas overflow from the contactor tank to the transformer tank;
- degass the oil;
- remove the transformer out for repair.

Each factor of this nature may cause an intense emission of methane. Discrete hotspots cause an increase in the concentration of ethane or ethylene, but if the basic gas is methane then this is caused by a cooling system defect or by personnel overloading the transformer. Numerous problems can be solved by adding heat sinks and fans or by changing transformer load.

2.1.1.2 Physical and Chemical Indicators for Estimating the Condition of Paper Insulation for Power Transformers in Operation

During the operation of a power transformer, the cellulose insulation of the windings is subject to destruction and dehydration processes aggravating its physical and chemical properties. First of all, this is reflected by deterioration of mechanical strength, oxidation, formation of pores, and chemisorption of acidic products due to the aging of transformer oil and metallic compounds of variable valence.

The electrical strength of oil-impregnated paper remains largely unaffected by aging, since the damaged areas of cellulose insulation are rapidly filled with oil, and the electrical indicators (insulation resistance and dielectric loss angle tangent) neither change significantly nor can serve as indicators of aging. Assessment of wear/aging for the winding insulation of any particular transformer should include a direct analysis of the physicochemical condition of cellulose insulation and of the related indicators for the degree of degradation in development. In addition, one needs a series of diagnostic features that allow one to make an objective conclusion on the amount of insulation wear and make a decision on the possibility and expediency of further transformer operation.

Due to a rather complex network of parallel and sequential chemical reactions leading to degradation, and the multiplicity of factors affecting the kinetics of their development, it is not possible to predict with the required accuracy the degree of deterioration of the winding insulation by analyzing the effects of operating factors. An assessment of deterioration for the winding insulation of any particular transformer should include a direct analysis for the physicochemical condition of cellulose insulation and for the related indications as to the amount of degradation development. It is evident that an acquisition of numerous diagnostic attributes for estimating the wear of winding insulation should be based on a profound study of physicochemical processes occurring in cellulose insulation under the influence of operational factors, among which the most significant ones are electric field, temperature, air (oxygen), chemically active impurities (aging products) and moisture.

The main physical and chemical phenomena causing the degradation of winding insulation during the operation of power transformers are as follows:

– thermal destruction and dehydration,
– hydrolysis of cellulose insulation,
– oxidative destruction when exposed to acidic products of oil aging and oxidants contained in oil,
– catalytic acid alcoholysis (alcoholysis is a general term for a group of exchange reactions between various organic compounds: epoxides, anhydrides and halides of carboxylic acids).

Electric field enhances the effect of most physicochemical factors and also promotes the adsorption (on the surface of cellulose insulation) of the aging products of transformer oil and construction materials. It also accelerates another important process of cellulose degradation, being catalytic acid alcoholysis under the action of hydroxyl-containing hydrocarbons (alcohols) in the presence of low molecular weight organic acids and other products formed in the oil during aging. The hydrolysis of cellulose insulation, which proceeds in parallel with the process of acid alcoholysis, makes, as compared to the latter, a substantially smaller contribution to overall degradation, which, in particular, is due to a rather low content of moisture in the insulation during normal transformer operation.

An important factor in the aging of cellulose insulation is its thermolysis, caused by elevated temperatures. Under the influence of high temperatures (more than 90 °C) in cellulose insulation, apart from an acceleration of the above-mentioned processes, thermal degradation is also activated, being destruction and dehydration in amorphous and mesomorphic regions with a formation of furfural and furan compounds. In addition, along with the indicated degradation processes, during transformer operation cellulose insulation is subject to oxidative destruction once exposed to acidic products of oil aging and to oxidants contained in these products. This process leads to the formation of oxidized (mainly carboxyl) groups and structure disturbances, as well as to the chemisorption of low molecular weight degradation products and acidic products of oil aging, copper and iron ions formed during the corrosion of metal elements of a transformer in the course of its operation. This process is accompanied by the release of carbon oxide and dioxide into the oil; a visual sign of catalytic thermal oxidative destruction of cellulose insulation of windings is its dark brown color.

In accordance with [16], two methods are provided for estimating the condition of winding paper insulation:

– according to the presence of furan compounds in oil;
– according to the polymerization degree of insulation samples.

It should be noted that the destruction of cellulose insulation during the operation of a transformer can be accompanied by a release of furan compounds into the transformer oil, the most of which should be regarded as furfural and hydroxymethylfurfural. Around 80% of furfural dissolves in insulating oil, and hydroxymethylfurfural is adsorbed mainly in paper insulation.

In accordance with [16], a permissible content of furan compounds (limiting the normal condition region) is set at no more than 0.0015% by weight (with furfural being no more than 0.001% by weight). For a number of reasons, discussed, e.g. in [1], this indicator does not reflect the actual degree of cellulose destruction, namely, the degradation dynamics for cellulose insulation. An objective indicator that makes it possible to assess the wear of winding insulation is the *degree of polymerization*, which directly characterizes the depth of its physicochemical destruction. At the same time, a decrease in the amount of polymerization has a monotonous dependence and reflects a monotonous decrease in the mechanical strength of insulation, which determines a deterministic diagnostic value of using this indicator. The resource of winding paper insulation is regarded to be exhausted when the degree of polymerization is reduced to 250 units, since in this case there is at least a 4-fold decrease in the mechanical strength of insulation in comparison with the original strength. This, in turn, dramatically increases the risk of short-circuited turns and transformer damage when mechanical forces make appearance, primarily in the case of through SC currents.

For an objective assessment of the wear sustained by transformer winding insulation, it is necessary to measure the degree of polymerization of a sample of turn insulation, selected in one of the upper coils [17]. The sampling of coil insulation can be carried out using a disconnected transformer both during overhaul and when partially draining the oil through the hatches.

Moisture contained (dissolved or bound) in transformer oil is one of the most important factors affecting the insulating properties of oil-paper insulation. The immediate reason for a decrease in the electrical density of transformer oil is the presence of moisture dissolved in it. However, bound moisture readily transforms into dissolved moisture, so that it is believed expedient to determine the total amount of moisture in transformer oil. Moisture content up to 200×10^{-6} rel. units hardly affects oil conductivity and dielectric strength. Once this quantity of moisture is exceeded, treelike structures of increased conductivity start to emerge, and then free water inclusions (drops) begin to appear, sharply reducing the value of electrical strength. Electrical effects and thermal aging of cellulosic material in electrical equipment lead to the appearance of carbon oxides and water, and therefore paper in electrical equipment is not only an adsorbent and principal moisture carrier, but also the main source of water.

Monitoring the process of *dehydration* due to polymerization is highly effective, since it allows one to examine the amount of wear sustained by the paper insulation of windings. Output of water released by paper with a polymerization degree of more than 300 rel. units is around 10^{-3}–10^{-2} % of the mass and has no significant influence on the performance of insulation. Upon reaching the degree of polymerization below 250 rel. units, the release of water due to dehydration may be more than 6% of the mass, which leads to a decrease in the dielectric strength of insulation [17].

Measurements of polymerization degree carried out to obtain a reliable assessment of the wear sustained by the winding insulation of power transformers should be made by determining the viscosity characteristics of cellulose insulation solutions in a cadmium-ethylenediamine complex. This makes it possible to ensure the absence of significant destructive changes in cellulose samples under examination, including oxidized ones. The use of other solvents tends to cause a chemical degradation of cellulose. Analyzing the polymerization of insulation by converting the latter into ethers may lead to overestimated values of the indicator in question, resulting from dissolution sustained by the low molecular weight fraction, with subsequent erroneous conclusions.

Mechanical impurities contained in oil dramatically reduce the dielectric strength of insulating gaps.

The electrical strength of the oil depends on the size of the impurity particles, since particles of larger size have a larger effect on increasing the field inhomogeneity in the oil, especially if they possess an increased conductivity, in particular due to the presence of moisture. Initially, transformer oil contains very small particles, caused by impurities in crude oil or formed during oil processing. During the manufacture and assembly of a transformer, its oil may be penetrated by cellulose fibers, dust, resin and metal particles. During the operation of a transformer, the concentration of particles, such as cellulose fibers, as well as metal and resin particles, increases as the materials age. The particles are carried throughout the entire transformer bulk by forced circulation of the oil. Local overheating and partial discharges also increase the concentration of carbon particles.

Having determined the nature of contaminants and aging products contained in transformer oil, one may proceed to choose optimal methods for regeneration of the used oils: for some oils a simple cleaning from mechanical impurities is sufficient, while for others deep processing is required, sometimes using chemical reagents. Used oil recovery methods can be divided into physical, physicochemical, chemical and combined ones. In practice, combined methods are normally used to obtain high quality recovered oils.

One particular case of oil contamination is the formation of sparingly soluble colloidal particles of copper and iron naphthenates. To detect colloidal particles contained in oil, turbidity measurements are used to determine the decrease in intensity due to the scattering by the particles of light passing through an oil cuvette.

The above makes it clear that the basic regeneration methods for used oils cannot be applied separately; in practice, it is often necessary to resort to various combinations of methods in order to ensure that the maximum cleaning effect is achieved. When choosing a regeneration method or a combination of methods, it is necessary to take into account the nature of aging products contained in used oils and the requirements for regenerated oils, as well as the amount of collected oils. One also needs to be aware of environmental consequences inherent in certain regeneration methods, so as to choose the most appropriate ones under the circumstances. Having these data available, one can determine the physicochemical properties of oil that need be corrected, and therefore one can choose an appropriate method of recovery.

The problem of removing moisture and impurities is effectively solved by the adsorption purification of oil using thermosyphon filters with an adsorbent, silica gel, due to its good adsorption properties and large specific surface area, being 350–450 m^2/g.

Dehumidified air needed for the operation of a transformer is obtained using air-drying cartridges filled with adsorbent (silica gel) and a small amount of silica gel indicator. The latter helps to determine the degree of adsorbent humidification and the need for its replacement or regeneration.

In the course of transformer overhaul, silica gel reduces the acid number of used transformer oils and gives them a second life. These processes of adsorption purification are extremely important in power engineering and can be achieved using sorbents of high and proven quality.

2.1.2   Vibration Control

The weakening of winding pressing can be regarded as the initial stage of damaged condition, causing the development of more serious and dangerous winding defects, such as coil descent and buckling, or turn-to-turn short circuits. Vibration control technology is used to monitor the degree of winding pressing. The purpose of transformer vibration diagnostics is to estimate the condition of a mechanical system, as well as to identify and eliminate defects in both external equipment (damage to pipelines due to resonant vibrations, wear of bearings of oil pumps and fans) and internal systems (unpressing of windings and magnetic conductor, displacement of magnetic shunts due to vibration.

Transformer vibrations are polyharmonic vibrations with frequencies being multiples of 100 Hz. The source of transformer vibrations is a magnetic conductor. The cause of this phenomenon is magnetostriction. Electric motors of oil pumps and fans are vibration sources by themselves, but their energy is much smaller. The vibration frequency of attached equipment is due to the rotational speed of electric motors, being approximately 720–1440 rpm. Vibrations are transmitted from sources to other units and elements of a transformer. An examination begins with measuring the vibration of a transformer tank. The most important vibration characteristics are as follows [18, 19]:

– vibration velocity, characterizing the vibration energy whose values are used to estimate the condition of the transformer tank, as well as to assess the impact of the transformer on the foundation;
– vibration acceleration, characterizing the inertial forces acting on the transformer tank as a result of the movement of its internal elements;
– vibration displacement, characterizing the vibration loads crucial for the condition of the transformer tank, welds and other elements.

The frequency spectrum of vibration velocity allows one to identify the vibration sources. Measurements are carried out in the frequency range up to 1000 Hz, since more than 90% of the total vibration energy of a transformer is concentrated in this range.

For a general assessment of transformer condition, a need for additional examinations arises at the following parameter values:

– vibration acceleration being more than 10 m/s^2,
– vibration speed being more than 20 mm/s,
– vibration displacement being more than 100 microns.

The condition of fans and oil pumps depends on the cooling system design. It can be estimated on a basis of the following criteria:

– indicator of defects in a blower fan is a vibration velocity of the bearings being higher than 7.1 mm/s;
– damaged condition of an oil pump is related to a vibration velocity being higher than 4.5 mm/s.

The pressing quality of the windings and magnetic conductor can be determined by examining the vibration spectrum on the surface of the transformer tank. Measurements are carried out in two modes, idle and loaded. It is assumed that in the idle mode the vibrations are caused by magnetostriction in the magnetic conductor, while in the loaded mode the additional effect of electromagnetic forces in the windings is taken into account. A decrease in the compression force of magnetic conductor sheets below a nominal value leads to an emergence of frequencies of 300–500–700 Hz. A weakening of winding pressing leads to a decrease in the 200 Hz frequency component.

The quality of winding pressing can also be estimated by measuring the natural frequencies of winding vibrations under mechanical shock. The basic method is registration of EMF induced on the windings at the bushings of a bus-bar transformer during a pulsed mechanical action. This process has the form of damped oscillations. The spectrum of EMF induced under pulsed mechanical action on transformer windings at different degrees of winding pressing is different and is shown in Fig. 2.1, see [20].

The disadvantages of this method are as follows:

– higher requirements for mounting a vibration sensor;

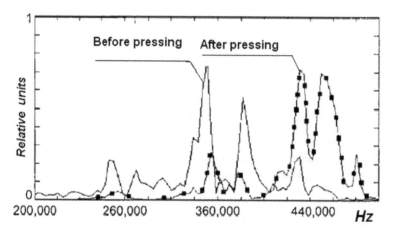

Fig. 2.1 EMF spectrum induced in the winding of a TTs-630000/500 transformer under pulsed mechanical action on the transformer winding at different degrees of winding pressing

– dependence of vibration parameters on a large number of factors, combined with the problem of isolating a vibration signal due to malfunction, which requires the use of methods of correlation and regression analysis.

Despite these disadvantages, the vibration control of the pressing degree in transformer windings is developing both in Russia [20, 21] and in other countries [22–25].

2.2 Electrical Control Methods

2.2.1 Insulation Condition Monitoring by PD Registration

Methods of PD registration are divided into two groups: non-electrical and electrical methods.

Non-electrical methods include the optical and acoustic methods.

The optical method is historically the first one and is based on a registration of PD glow. The method allows one to control PDs at electrode edges and in transparent materials. PD control inside opaque structures is impossible, which puts significant limits on its use in the diagnosis of high-voltage transformers, whose insulation is not transparent optically.

The acoustic method makes it possible to register PDs inside opaque objects. The apparent simplicity of the method does not compensate for its massive problems in determining the location of PD occurrence in high-voltage transformers. To detect such occurrences, supersensitive microphones are utilized, which detect sound waves located in the frequency range above the audibility threshold. Research on the improvement of the method is in progress.

The sensitivity of **electrical methods** is higher than that of non-electric ones, so they are more widely used at present. They allow one to detect the largest number of PD characteristics and use them to estimate the condition of transformers. Most electrical methods do not need the object under analysis to be supplied by electrical voltage being much higher than the rated operating values, so they are gentle on the insulation of electrical equipment. However, many of these diagnostic methods require the measuring instruments to be in contact with the object being diagnosed, which does not contribute to simplicity and convenience. Due to the high sensitivity of the methods, it is necessary to take a series of measures and use special equipment for debugging induced interference.

The electrical methods are divided into three main types:

(1) *Indirect methods of PD registration*
(2) *Electromagnetic or remote (PD registration with antennas)*
(3) *Registration of high-frequency (HF) electromagnetic radiation generated by PDs.*

(1) *Indirect methods of PD registration* include those that allow one to determine dielectric losses by measuring the tangent of dielectric loss angle in insulation (tgδ), or by measuring Volt–Coulomb characteristics. These methods provide an idea about the voltage of PD occurrence judging by a sharp increase in tgδ. Since the application of these methods involves a summation of different types of losses, it is difficult to single out the direct losses due to PDs. In addition, these methods have low sensitivity.

A practical implementation of this idea is the *control of an ionization curve when measuring tgδ*. The method is based on the fact that PD occurrence at an insulation defect causes a decrease in its resistance, and thereby an increase in active current (and hence tgδ); the reactive current is almost unchanged. The voltage dependence of tgδ is determined using a high-voltage Schering bridge. The voltage at which ionization starts in an insulating structure corresponds to the onset of growth in tgδ, and the inflection point of the curve is regarded as the onset voltage of ionization, Fig. 2.2.

The drawback of the method is that the recorded increase in tgδ is related not to the onset of discharges, but rather to an intense process of PDs repeated numerous times during the cycle, which covers a significant part of foreign inclusions in the dielectric volume. For this reason, the method can be classified as obsolete.

(2) *Electromagnetic or remote method (PD registration with antennas).*

This method is remote and allows one to detect PDs using a directional receiving microwave antenna device. The use of this equipment does not depend on the voltage class, which is an advantage. The disadvantage of this method is the inability to obtain any quantitative information about the processes and phenomena accompanying PDs,

Fig. 2.2 Curve for
determining the onset of an
ionization process

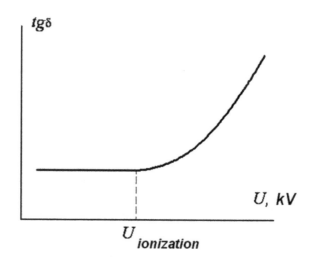

as well as the effect caused in the readings of a receiving device by electromagnetic
radiation from external sources.

Research is currently in progress to improve the adaptation of the method to opera-
tional requirements. Even now the method allows one to carry out lasting continuous
monitoring of equipment accompanied by information transfer to an operator with
the aid of digital technologies.

The most active research on PDs in isolation began in the late 1950s and early
1960s. At that time, the destructive action of PDs was first encountered in a large
group of ultra-high voltage equipment in the late 1950s during the operation of
400 kV transformers, and later on, 500 kV ones.

Since the 1980s, the strategy for diagnosing equipment in Europe and America
has gradually changed: there has been a shift from the concept of routine testing to
the concept of testing determined by estimating the technical condition of equip-
ment. It is known from practice that positive results of testing equipment by high-
voltage, which has been currently regulated in Russia, do not guarantee protracted
trouble-free operation of the tested equipment. Besides, during the insulation test
the condition of equipment significantly deteriorates due to the supply of voltages
being 4–6 times higher than the nominal value. The diagnostic methods for PD
registration, at the same time, allow one to give the most accurate estimation of the
residual life of equipment, while not affecting the isolation in any significant way.
This is ensured by applying lower voltages to an inspected object, in some cases
being close or equal to the nominal value. In parallel, the task of creating a system
of automated monitoring of equipment parameters under operating voltage has been
undertaken. The largest success in this direction has been achieved by High Voltage
Partial Discharge ltd (HVPD) that producing devices for on-line PD monitoring in
the insulation of electrical equipment. To date, this is almost the only method that
permits a detection of incipient local defects in the process of insulation degradation.
A wide application of the method is constrained by the complexity of measurement

techniques, the high costs of equipment, and the high requirements to qualifications of the personnel operating the equipment.

(3)   *Registration of electromagnetic PD-generated HF radiation* has become the most widespread technique, since it allows for reliable measurements of basic PD characteristics and provides high sensitivity.

The method of registering high-frequency PD components, implemented using partial-discharge indicators (PDI), is also widely used.

The efforts of researchers and engineers are aimed at creating instruments and techniques that should make it possible to register the weakest PDs and provide measurements of energy dissipated by single discharges. This requirement has now been met by using PDIs connected directly to a discharge circuit. Such indicators consist of a receiving circuit, an amplifier and a measuring device. This is based on the measurement of an apparent charge:

$$\Delta Q_0 = \Delta U C_0, \tag{2.2}$$

where C_0 is the insulation capacitance.

Ripples of voltage ΔU are measured and fed through an amplifier to an oscilloscope. The moment the pulses appear on the screen determines the voltage of ionization occurrence, while the amplitude and frequency of the ripples determine the PD intensity. There are several options for this scheme. Two of them are shown in Figs. 2.3 and 2.4.

(a)   An active resistance scheme is shown in Fig. 2.3.

Resistance R_1 is connected in series with a tested object and a voltage drop on the object is recorded by a PDI. The resulting oscillogram is interpreted as to the presence or absence of PD. The disadvantage of this method is its low noise immunity.

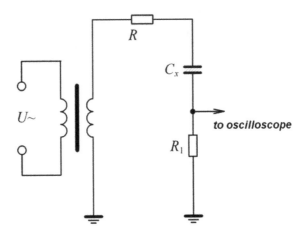

Fig. 2.3 Active resistance scheme for PD detection: R is a protective resistance; C_x is a tested object; R_1 is an active resistance

$$U_1 < U_2$$

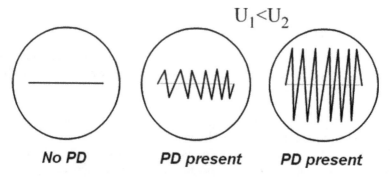

No PD   PD present   PD present

Fig. 2.4 Oscillograms for different PD intensities

On the oscilloscope, one can observe the picture shown in Fig. 2.4.

(b)   A scheme with inductance and capacitance is shown in Fig. 2.5.

The PDI is connected to the tested object through the dividing capacitor C_{div}, which serves as a blocking filter for currents of the operating frequency. At the emergence of PDs in the object (C_x), chaotic voltage fluctuations in the object excite continuous periodic oscillations in the PDI with a frequency corresponding to the oscillation period of the circuit:

$$T = 2\pi\sqrt{LC}. \tag{2.3}$$

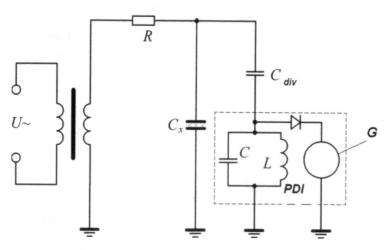

Fig. 2.5 PD detection scheme with an oscillatory circuit and a galvanometer: R is a protective resistance; C_x is a test object; C_{div} is a dividing capacitance; L–C is the oscillatory circuit; G is the galvanometer; PDI is a partial discharge indicator

The order of the PDI tuning frequency is usually chosen as several tens of kilohertz. The apparent intensity of ionization is determined using the value of ΔU given by formula (2.2).

In conclusion, it should be noted that the use of the PD method for preventive tests is very promising, since it allows for continuous monitoring under operating voltage. Unfortunately, it also has some drawbacks that demand it to be used in combination with some other methods:

(1) The presence of a large amount of interference that makes it problematic to decipher the results obtained (the source of interference may be the appearance of a corona on the wires, the sparking of the collectors of electrical machines, etc.);

(2) Rather than the presence of failures, the method detects the presence of PDs, while there may well exist dangerous defects without any PD involvement: cracks filled with water or some other conductive liquid, carburized pores in which PDs have stopped.

Technical implementations of the concept of insulation diagnostics by PD registering are extremely diverse.

Measurement of PD electrical characteristics and determination of PD origin location are successfully implemented by many high-voltage laboratories at large energy companies in the USA, Australia, Japan, Germany, China, Canada, Brazil, as well as transnational energy companies EDF and Enel, and others, where this control method is included in the mandatory insulation monitoring programs. Preference is given to automated systems and devices operating in real time. The principal contribution to these developments is due to such large companies as ABB, Siemens, Mitsubishi, Tettex, General Electric, Megger, Omicron, Haefely. To monitor the insulation condition of a power transformer of the highest voltage class (1000 kV), Mitsubishi has developed a highly sensitive PD monitoring system.

Tettex Instruments AG produces the multichannel PD analyzer Digital Partial Discharge Detector DDX9101, a digital meter with computer data processing. The software of this device allows displaying PDs both as Lissajous figures (with reference to the phase of the operating voltage) and as three-dimensional images showing the amplitude-frequency characteristics of discharges [26] (Figs. 2.6 and 2.7).

New opportunities have appeared for monitoring defects in insulation when measuring PDs on a basis of digital equipment. An example of such equipment is a partial discharge analyzer supplied to the world market by Haefely and Tettex.

Fig. 2.6 Digital Partial Discharge Detector (DDX9101)

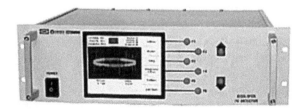

Fig. 2.7 The DDX 9121b is the latest in the DDX family of partial discharge (PD) & radio interference voltage (RIV) testing instruments. It is a fully digital state-of-the-art high-performance PD detector [27]

Anti-interference is one of the most challenging tasks in detecting PDs and measuring their parameters in a production environment. Research is underway to overcome difficulties of PD measurements in operating transformers under the conditions of operating substations. For example, the University of Hanover (Germany) has developed a computer-aided system of filtering signals from interference. Acceptable signal-to-noise ratios in operating substation conditions have been achieved through the use of digital filters, adaptive limiting filters, and measurement circuits using the opposite directionality of PD signals from the transformer and that of noise coming from the power grid.

Researchers and engineers at **OMICRON** (Austria) have a large experience in digital PD analysis of high-voltage structures. The research center of the company maintains a database on the results of PD measurements in transformers and other high-voltage equipment. The analysis exploits the theory of pattern recognition and the three-dimensional spectra of PD distribution by repetition frequency, as well as by amplitude and location relative to the test voltage phase. One of the latest developments is a portable system for measuring and periodic monitoring of partial discharges in diverse high-voltage equipment, OMS 605, see Fig. 2.8.

The OMS 605 is a portable system for measuring and online monitoring of partial discharges on equipment in operation according to IEC 60270. In contrast to offline diagnostic tests, online PD monitoring ensures the assessment of insulation condition during the operation of equipment. Data on the trends in insulation condition allow one to estimate the rate of wear and tear. This is critical information that helps one optimize the equipment maintenance schedule and maximize the return on investment.

The system is used to periodically assess the insulation condition of medium and high voltage equipment under operational load. All the necessary devices for diagnostics and monitoring of partial discharges are housed in a sturdy case with wheels, which is easy to transport from initial place to other one. The OMS 605 system is used with a range of capacitive and inductive PD sensors to cover the entire frequency range in which frequency discharges can be detected, including the

Fig. 2.8 Portable system for measuring and periodic monitoring of partial discharges in diverse high-voltage equipment, Omicron OMS-605

microwave range. These sensors can be permanently installed, so that frequently tested equipment is not taken out of service each time they are installed.

Thanks to a synchronous data collection using three channels, the operator receives a comprehensive set of PD information. A fourth channel can be added for an optional PD sensor or strobe. Advanced noise suppression and source separation techniques enable accurate partial discharge location. The OMS 605 software visualizes the acquired data, as well as allows one to record data streams in real time and save them for later analysis. The OMS 605 can be used for periodical monitoring the status of diverse high-voltage equipment. One portable device can be used to monitor different objects. A convenient auto-tuning system and versatile software make it easy to adapt the OMS 605 for monitoring any type of high-voltage equipment. The features of the OMS 605 system are as follows:

- Ease of transportation for periodic monitoring of various objects;
- Easy-to-use self-adjusting modules for quick system preparation for work;
- Synchronous data collection through several channels for a comprehensive estimation of PD activity;
- Compatibility with various PD sensors;
- Rugged and reliable industrial grade design (IP65);
- Automated noise suppression and source separation for accurate PD detection;
- Powerful software for analysis and visual display of PD data.

The system is used by personnel specialized in testing and maintenance of companies, being equipment manufacturers, network and industrial enterprises, power plants, as well as by personnel of service companies for periodic inspection and monitoring of equipment for PD presence and subsequent assessment of the insulation condition of power transformers of all voltage classes.

Another example of OMICRON's productive creativity is the Omicron MPD 800 universal partial discharge measurement and analysis system, Fig. 2.9.

Fig. 2.9 Omicron MPD 800 universal partial discharge measurement and analysis system

The MPD 800 is used during standard-compliant PD testing for routine and type tests, factory and on-site acceptance tests, as well as for troubleshooting aimed to localize or analyze PD sources in all types of high voltage equipment.

The MPD 800 system consists of an MPD 800 measuring device, an MCU2 control unit, and an MPD Suite of software. Depending on the measurement, the MCU2 connects to one or more MPD 800 devices via fiber optic technology. The MPD 800 and RBP1 batteries are connected to a test object directly or via CPL1 or CPL2 couplers. MCU2 connects via USB to a laptop or PC with MPD software installed for analysis. This approach has several advantages:

- Safe approach to testing due to galvanic isolation
- Battery powered
- Minimum environmental impact
- High synchronicity for improved partial discharge analysis

Among other companies involved in the study of electro-physical PD processes and technical means of their control, the companies *MEGGER* and *ABB* stand out. The companies use a continuous PD registration system using digital technology. Based on the study of relations between the measurements produced by electric and acoustic sensors, a technique for identifying the location of PD occurrence has been developed.

MEGGER's engineers have developed the *Power Diagnostix ICM compact*, a standalone PD measurement tool, shown in Fig. 2.10.

The Power Diagnostix ICM compact has a simple push-button interface and on-screen menus on a built-in liquid crystal (LCD) panel. The LCD display modes include a simple PD meter with adjustable "needle" sensitivity, phase-resolved monochrome PD charts for defect classification, and an oscilloscope-type display showing phase-summed charge pulses superimposed on the applied voltage wave. Although the ICM compact is a stand-alone device, it can be connected to a computer running Power Diagnostix software to take screenshots or remotely control the device.

By instantly displaying information in an intuitive interface, the ICM compact is a good choice for applications such as proof-testing in the manufacture of electrical products, and such as quality assurance in municipal and industrial equipment, starting from capacitors and bushings to gas-insulated switchgears, voltage transformers, etc.

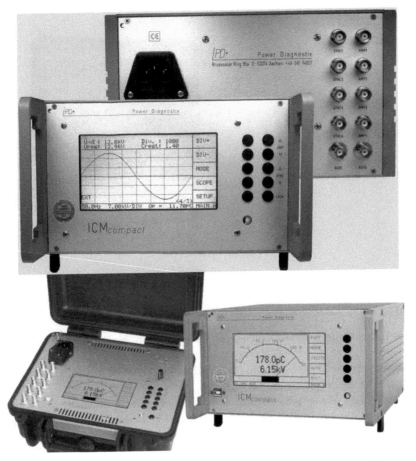

Fig. 2.10 Autonomous complex for measuring partial discharges *Power Diagnostix ICM compact*

To adapt the ICM compact base unit to specific measurement requirements, it can be equipped with various options:

- Locating cable damage.
- Analog strobing to eliminate external noise. The option offers sensitive measurements even in noisy environments.
- MUX4, a four-channel multiplexer for testing three-phase equipment such as power transformers. The device supports individual configurations and calibrations for each channel.
- MUX12. The option provides a built-in 12-channel multiplexer or a remote 12-channel breakout box for acceptance testing of large power transformers.
- AUX4. Provides recording of up to four additional parameters as 0 (4)-20 mA or 0–10 V signals for long-term testing.

Fig. 2.11 ABB
Ability—SWICOM
switchgear condition
monitoring device

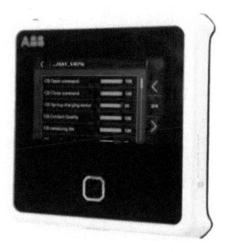

- RIV measurement. Adds a function of measuring the strength of radio emission to the device.
- Modem. Provides communication with the device via an analog telephone line.

At the same time, measurements can be made using both acoustic and antenna methods.

Among the ABB developments, the ABB Ability—SWICOM switchgear condition monitoring device stands out (Fig. 2.11).

The ABB Ability—SWICOM switchgear condition monitor with partial discharge detection is a cost-effective solution for monitoring equipment condition. SWICOM is a diagnostic tool that monitors the running order of the entire stock of mechanical and electrical equipment. The device collects data from IEC 61850 compliant protection relays, as well as additional sensors, and converts it into diagnostic information. It can be installed on virtually any new or existing air or gas-insulated switchgear of medium voltage. Due to the high cost of efficient and reliable equipment, it was only high-voltage switchgear that had been tested for such discharges earlier, and the equipment had to be switched off to examine the insulation. The ABB Ability—SWICOM switchgear condition monitoring device with PDCOM sensors compensates for this disadvantage and represents an effective solution for monitoring the condition of medium-voltage equipment.

The SWICOM system has a number of advantages:

- Monitoring the circuit breaker drive, the temperature at critical points in the primary circuit, and the partial discharge using an indicator.
- Complete retrofitting of any existing panel, regardless of age, design features or device type, while ensuring compatibility with ABB digital.
- Fast and easy integration into truly digital switchgear, also when the panel design is not specified for this application.

Acoustic methods, despite their inherent disadvantages, have found extensive use in PD detection and location during measurement on operating equipment due to their lower sensitivity to extraneous electrical signals (interference). An acoustic assessment of PD intensity is recommended as part of diagnosing the condition of power transformers. Particularly effective are acoustic PD sensors placed inside the transformer tank. Such sensors are used to control of particularly critical transformers and also in type tests.

Research on acoustic waves arising from discharges in oil has been conducted by a number of scientific institutions in Japan. Techniques for locating the PD site in a transformer have also been developed. In particular, a 1000 kV high-voltage power transformer produced by Mitsubishi is equipped with acoustic sensors in the tank.

At the Polytechnic Institute in Poznan (Poland), the mechanism of sound wave propagation from a PD source has been studied in detail. An innovative technique for measuring and locating the PD site in a transformer has been developed using an acoustic method. The result of PD registration with the site localization is shown in Fig. 2.12.

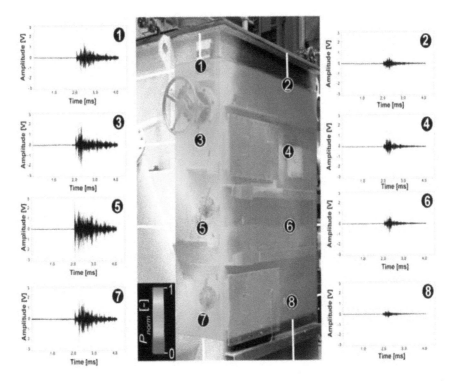

Fig. 2.12   The result of partial discharge source location with advanced auscultatory technique and example of AE waveforms recorded in 8 locations on the transformer tank [28]

Acoustic PD sensors, both inside and outside the tank, have been part of the EPRI control system for testing a 525/345 kV, 300 MVA transformer at the Ramapo substation in the ConEd electric power system.

The PTCSM automated system for monitoring the condition of 400/161/161 kV transformers with a power of 195/97/97 MVA of the FennoScan DC cable line, operated since 1988, is equipped with acoustic sensors for PD detection.

PD sensors (electrical and acoustic) are included in automated systems for a continuous monitoring the condition of large power transformers, developed by Westinghouse, Siemens, ABB, see [29].

In Russia, the most profound issues of monitoring PDs in electrical equipment under operating voltage have been worked out by Dimrus Company and by Siberian Research Institute of Power Engineering (SRIPE). The researchers of SRIPE regard the control of PDs in combination with CADG as a basis for transformer diagnostics in operating mode.

Monitoring the condition of transformers by PD characteristics using electrical and acoustic sensors is widely used in the power systems of Russia. As an example, we can name some Russian enterprises that successfully use such diagnostics: Electrosetservice, Technoservice-Electro, JSC "Firm ORGRES". They use a technique well-protected against interference for acoustic measurements and PD location in power transformers. Among the PD sensors of the latest generation, one can point out indicator of PD characteristics with Russian name IChR-201, Fig. 2.13.

The main technical parameters of the IChR-201 device are given in Table 2.3.

The IChR-201 device implements all the recommended by State Standard 20074 and IEC 60270 schemes for measuring PD characteristics. The device simultaneously measures in real time the values of all the PD characteristics recommended by the standards:

- Maximum apparent charge of PD pulses (according to State standard) 20074 or IEC 6028);

Fig. 2.13 Partial discharge indicator IChR 201

Table 2.3 Main technical parameters of IChR 201

Parameters and characteristics	Value
Test voltage frequency, Hz	40–400
Sensitivity (when a signal is applied directly to the device input), pC, no worse	0,1
Minimum measurable apparent PD charge, pC, no more	1,0
Maximum measurable apparent charge of PD, pC	10000
Measurement error of PD charges in the range of 1–10 pC, pC, no more	±1,0
Measurement error of PD charges in the range of 11–10000 pC, pC, no more	±10
The highest repetition rate of measured PD pulses, kHz, no	100
Bandwidth limits, kHz	45–700
Pulse resolution time, μs, no more	6,0
Digitization of the signal, bit	14
Test voltage input characteristics: – input impedance, Mohm, no less – input capacitance, pF, no more – maximum voltage of frequency 40–400 Hz, V, no more	1,0 50 100
Supply voltage (50/60 Hz), V	220 ± 10%
Power consumption, VA, no more	60
Dimensions, mm	110 × 170 × 250
Weight of measuring units (excluding cables and computer), kg	6,0
Packing dimensions, mm (LxWxH)	552 × 250 × 430
Weight of the complete set in the package (including cables, computer, packaging, documentation), kg	23

- Repetition rate of PD pulses;
- Average current of PD pulses;
- Pulse power of PDs;
- Root-mean-square parameter.

The apparatus provides real-time visual display of various (at the option of an operator) types of measurement oscillograms, including three-dimensional ones, as well as the graphs of a test in progress with PD characteristics measurements. The device software (PDScanner 2.0) allows the operator to select the modes and ranges of PD characteristics measurements in the object, to calibrate the measurement scheme (in manual or automatic mode), and also to save the results of calibration and measurements as test reports automatically generated by the device. In the course of measurements, the operator can record any moment of measurement in a protocol as an oscillogram with all the PD characteristics measured at that moment. The noise, threshold and position filtering of a measured signal executed by the device ensure a high noise immunity of measurements. The device, as a rule, does not require any special, carefully shielded chambers: it is successfully used in the production of electrical equipment.

Electrosetservice uses a system for diagnostics and isolation of power transformers developed by the Special Design and Technology Bureau of the Mosenergo Association. The system detects the presence of PDs by an acoustic method in the frequency range of 40–100 kHz; it has 20 acoustic sensors and a stationary recorder unit. To determine the coordinates of the location of discharges according to the Mosenergo method, the sensors are placed in the form of a three-pointed star on the surface of the tank, the location being determined by the difference in the time of signals arrival. The intensity of discharges is determined by the magnitude of the signals, taking into account the depth of the occurrence of the site of discharges.

The Research and Development Center ZTZ-Service, which has performed surveys of many transformers in Russian power grids, considers the detection of PDs as a promising technique for monitoring the insulation condition of power transformers. At the same time, according to the experts, the difficulty of protecting against interference during measurements poses a serious problem.

DIMRUS stands out amongst Russian developments and equipment manufacturers. The appearance of one of its latest developments is shown in Fig. 2.14 (Tables 2.4 and 2.5).

The operation principle of the meters is based on a conversion technique using an analog-to-digital converter of information about PD signals, which is extracted by sensors included in the meters. There are two basic modes of operation:

– mode of periodic measurements of the maximum voltage amplitude and the number of PDs per one second,
– "temporary" monitoring mode [30].

Fig. 2.14 R2200 designed by *DIMRUS*

Table 2.4 R2200 parameters

No.	Parameter	Value
1	Number of channels for PD measurement	9
2	Frequency range of measured discharges	0,5–15,0 MHz
3	Dynamic range of registered PD	70 dB
4	Synchronization of PD pulses registration	Internal–External
5	Color screen resolution of the device, points	480*640
6	Computer interface	USB
7	Operating temperature range of the device	$-20 + 40$ C
8	Operating time powered by the built-in battery, h	5
9	Device weight without sensors, kg	3,5
10	Overall dimensions of the transport case, mm	520*430*220
11	Kit weight in a transport case, kg	21,5

Table 2.5 AR200 device parameters

No.	Parameter	Value
1	Number of channels for PD measurement	1 acoustic
2	Frequency range of measured discharges	30 kHz–300 kHz
3	Data presentation	LCD backlit display
4	Color screen resolution of the device, points	128*64
5	Standby mode	At least 15 h
6	Overall dimensions of the transport case, mm	160*120*38
7	Device weight, kg	1

To register PDs acoustically, **DIMRUS** has developed an AR 200 sensor (Fig. 2.15).

The apparatus for detecting PDs acoustically consists of a primary transducer (sensor) and a measuring device. The acoustic sensor converts a pressure pulse into an electrical signal. Piezoelectric transducers are commonly used. During operational inspection, the sensors are placed on the surface of an oil-filled transformer tank. This device has a significant drawback: it does not record low-intensity PDs (Fig. 2.16).

A similar device UltraTest has been developed for registering PD in the ultrasonic frequency range. The appearance of the device is shown in Fig. 2.12 [30].

To date, several models of devices and system architectures have been developed and used for detecting and measuring PD parameters by using their electromagnetic radiation. For example, Fig. 2.17 shows a directional antenna device with an HVPD Longshot operating unit and an HFCT sensor. The system allows detecting PDs by oscillations of electromagnetic waves in the frequency range from 300 MHz to 3 GHz. Its disadvantage is the impossibility of controlling the intensity of the discharges.

Fig. 2.15 AR 200 sensor

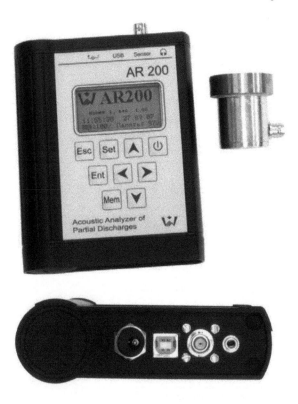

TEV (Transient Earth Voltage) sensors, which are capacitively coupled devices, are effective for measuring local PDs in insulation. For PD measurements, a TEV sensor must be fixed to a grounded metal surface (Figs. 2.18 and 2.19).

Among the promising methods of PD detection, a technique based on a registration of ultrasonic PD vibrations stands out, using sensor detectors and data analysis through wireless communication in ON-LINE mode. A technology based on the acoustic method is being developed in China [31]. The concept of the method implementation is shown in Fig. 2.20.

High-performance sensor detectors located at a certain angle in relation to the equipment and to each other capture the slightest ultrasonic waves and transmit the data to a cloud server, where they are analyzed and decoded in detail.

An effective system for analyzing PD pulses in the time and frequency ranges has been proposed in Switzerland.

The correct analysis of PD-current impulses in the time and frequency-domain requires wideband-coupling devices directly connected to the terminals of the power transformer. High frequency current transformers (HFCT) with a frequency range between 0.1 to 30 MHz have been successfully applied for this purpose [32].

To provide multi-terminal detection capability, which is important for the PD-analysis in the complex insulation systems of transformers or generators, the HF-CT's are directly connected to the capacitive tap of each transformer bushing (including

Fig. 2.16 Portable device
for acoustic PD registration
in the ultrasonic range

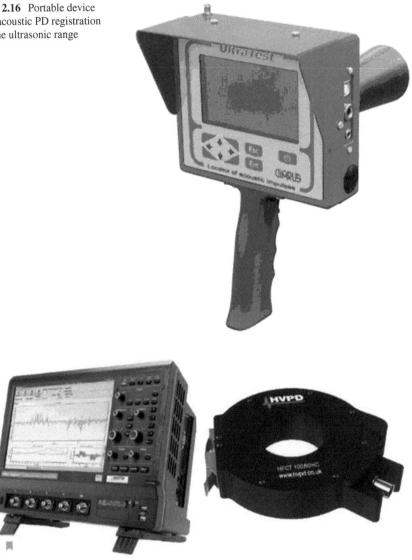

Fig. 2.17 HVPD antenna device by Longshot

low voltage bushings and neutral terminal bushing), or to the external coupling capacitors at the generator terminals. Advanced PD-detection systems (see Fig. 2.21) were introduced to analyse both, PD-pattern (statistical analysis of the PD-signal magnitudes) and PD current impulses in the time- and frequency-domain [33].

The main components of the advanced PD-detection system consist of commercially available instruments: (1) Spectrum analyser with a typical frequency span

Fig. 2.18 HVPD Longshot TEV sensor

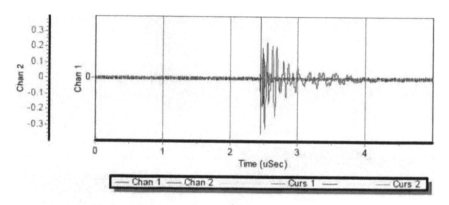

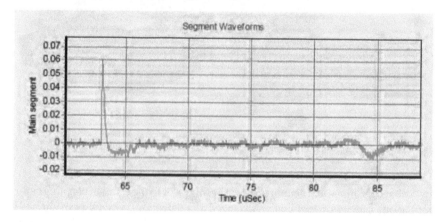

Fig. 2.19 PD pulses measured with a TEV sensor

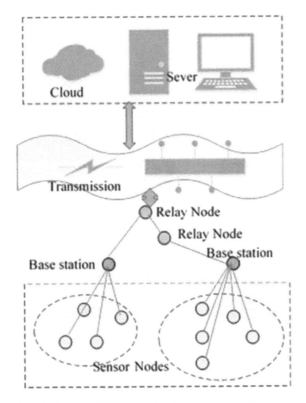

Fig. 2.20 PD registration based on highly sensitive detectors and wireless communications

between 1 kHz and 20 MHz. The spectrum analyser can be used for the analysis of PD-signals in the frequency-domain (full span mode) or for the quasi-integration of PD-signals to evaluate the value of the apparent charge using the variable band-pass filter (zero span mode). Normally the band pass filter of the spectrum analyser is utilized as a front-end of a PRPDA-system to record PD-pattern. (2) The multichannel digital oscilloscope is an analyzing device for PD-signals in the time-domain. The oscilloscope must have at least two channels with a bandwidth larger than 500 MHz. (3) The PD-analyser (PRPDA-system) is a computer controlled impulse acquisition and digital signal processing system preforming statistical analysis of recorded PD-data (PD-pattern). In a typical measuring set-up for power transformers or power generators, multi-channel PD analysers are used to record and process the signals from all terminals simultaneously.

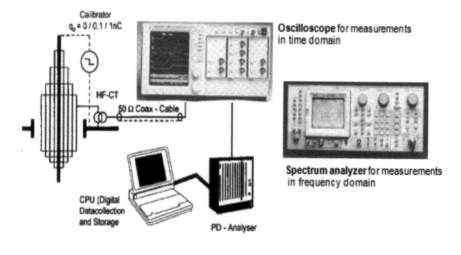

Fig. 2.21 Advanced PD-detection system: (1) spectrum analyzer, (2) digital oscilloscope, (3) PD-analyser (PRPDA system, (4) wide band current transformer (HF-CT) connected to bushing tap

2.2.2 Idling Loss Measurement

The losses in a power transformer are known to consist of so-called copper and steel losses. Copper losses are associated with a flow of load current through winding conductors, having a certain electrical resistance. Losses in the steel of a core are due to eddy currents, being magnetizing currents arising in a magnetic conductor.

According to the Rules for Electrical Installations (REI), measurements are made for transformers with a capacity of 1000 kVA and higher, at a voltage on the LV winding equal to that specified in the factory test report (passport), but not higher than 380 V. Idling losses of three-phase transformers are measured at a single-phase excitation according to the manufacturer scheme. When commissioning a transformer, the ratio of losses on the different phases of three-phase transformers should not differ from the factory data by more than 5%, and for single-phase transformers, the difference of the measured of losses from the original ones should not exceed 10%.

An open-circuit test for single-phase transformers is made at a voltage of 380 (220) V or less, supplied from the LV side at a frequency of 50 Hz in accordance with Fig. 2.22a, b. The supplied voltage must not exceed the rated voltage.

The applied voltage, the current, and the power P consumed by the tested transformer and the measuring devices are measured in accordance with Fig. 2.22a. Then the power consumed by the measuring devices $\sum P_{MES,DEV}$. is measured in accordance with Fig. 2.22a, b.

Losses in the transformer (P_0) are calculated using the formula

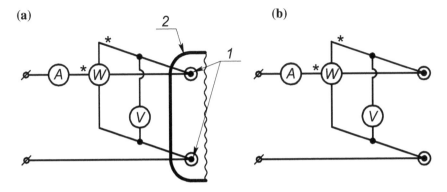

(a)   **(b)**

Fig. 2.22 Scheme for measuring the losses of a single-phase transformer XX: **a** measuring the losses of a tested transformer, **b** measuring the losses of measuring instruments: A—ammeter, V—voltmeter, W—wattmeter

$$P_0 = p - \sum P_{MES,DEV}. \tag{2.4}$$

Note. In the circuits of Fig. 2.22, it is allowed to use current-measuring instrument transformers.

Experiment of a three-phase transformer at low voltage is carried out in as the following three single-phase experiments made in accordance with experiment for single-phase transformers.

The first experiment: the windings of phase A are short-circuited; phases B and C of the transformer are excited and the losses are measured.

The second experiment: the windings of phase B are short-circuited; phases A and C of the transformer are excited and the losses are measured.

The third experiment: the windings of phase C are short-circuited, phases A and B of the transformer are excited and the losses are measured.

Short-circuiting of any phase is made at the corresponding terminals of any of the transformer windings (high, medium or low voltage).

Any of the transformer windings can be used to short-circuit one phase. The connection diagrams of transformer windings are shown in Fig. 2.23.

A measurement is usually made by applying voltage to two of the phases and short-circuiting the third phase of the low voltage (LV) winding, which results in more excitation of the magnetic system. The lower voltage makes it easier to measure voltage, power and current, therefore it is common to conduct experiment from the side of the LV winding.

The disadvantage of this method is the limited accuracy and reliability of identifying the winding defects.

Fig. 2.23 Connection
diagrams for the windings of
a three-phase transformer

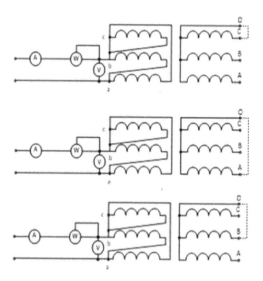

2.2.3 Transformation Ratio Measurement

The transformation ratio (TR) of power transformers is tested in order to confirm the compliance of the actual TR with the nominal one. TR is determined on all the winding branches for all the phases; on those winding branches that are inaccessible for switching on an assembled transformer this is determined before the complete assembly of a transformer. When testing three-winding transformers and transformers with split windings, TR may be examined for two pairs of windings, and measurement on all the branches of each of the windings is sufficient to be conducted once.

If a tap changer of winding branches has a tap selector, which is used to reverse the adjusting part of a winding or switch the coarse-control steps, then measurements can be carried out with one position of a tap selector corresponding to the lowest of the voltage values on the regulated winding. In addition, one more measurement is carried out in all other positions of the tap selector.

When testing three-phase transformers with three-phase excitation, line voltages are measured, corresponding to the same line terminals of the tested windings. If it is possible to measure phase voltages, it is allowed to determine the transformation ratio by the phase voltages of the corresponding phases. The transformation ratio for phase voltages is checked with single-phase or three-phase excitation of the transformer.

When testing three-phase transformers with windings connected according to the "star-delta" and "delta-star" circuits, the transformation ratio for phase voltages is determined with an alternated short-circuiting of phases. In this case, one of the phases connected in a "triangle" (for example, phase A) is closed, then, with single-phase excitation of the linear ends, the transformation ratio of the remaining free pair of phases is determined, which, in this method, should be equal to $2\,K_{ph}$ (if the HV

winding is connected as a "star") and 0.5 $\mathrm{K_{ph}}$ (if the LV winding is connected as a "star"), where $\mathrm{K_{ph}}$ is the phase transformation ratio. Measurements are carried out in the same way with short-circuiting phases B and C.

When testing transformers with the same winding connection schemes, it is allowed to carry out measurements with three-phase excitation, if it is established that the difference between the highest and lowest line voltages does not exceed 2%.

The transformation ratio is determined using two voltmeters. Voltage is applied to one of the transformer windings and is measured with one of the voltmeters. At the same time, the voltage on the other winding of the transformer is measured with the other voltmeter.

It is allowed to use measuring voltage transformers, as well as external resistors additional to voltmeters. The accuracy class of voltage transformers and additional resistors must be at least 0.2. The supplied voltage must not exceed the rated voltage of the transformer, but must be no less than 1% of the rated voltage. It is allowed to supply a voltage of less than 1% of the rated value if a voltage transformer is required when supplying a voltage exceeding 1% of the rated value. A voltmeter on the side of supplied voltage is allowed to be connected to the supply wires if this has no significant effect on the measurement accuracy. When measuring the transformation ratio, the resistance of the measuring circuit wires should be no more than 0.001 of the internal resistance of the voltmeter.

Conclusions on the consistency or inconsistency of measurement results are made on the basis of analysis for the measured value of transformation ratio.

The main disadvantage of the method is a limited number of identifiable defects, having a high degree of development.

2.2.4 Insulation Resistance Monitoring for Transformer Windings

The insulation resistance measurement method is the simplest and most accessible one. It is based on the characteristics of a change in the electric current passing through the insulation after applying a constant voltage to it. The insulation of transformer windings is a non-uniform dielectric. In the general case, an electrical equivalent circuit of insulation can be represented in the form of three branches (Fig. 2.24): the geometric capacitance of the dielectric C_1, the geometric capacitance of the dielectric layers C charged through the resistance of the dielectric layers R, and the insulation resistance of the dielectric R_1.

When a constant voltage is applied to the terminals of a circuit, the current will consist of an arithmetic sum of three components:

(1)   capacitive current I_r, due to the so-called geometric capacitance C_r; the current I_r almost instantly drops to 0, since the capacitance C_r is connected to a source without resistance and does not affect the results of measuring the resistances R_{15} and R_{60};

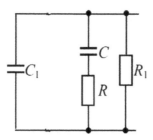

Fig. 2.24 Isolation-equivalent circuit

(2)   absorption current I_{abs} flowing along the R_{abs}–C_{abs} branch; this current reflects the process of charging a number of dielectric layers through the resistance of the previous layer. When the insulation is moistened, the resistance R_{abs} decreases, and the capacitance of C_{abs} increases, and therefore, for a more humidified insulation, the current I_{abs} has a greater value and drops faster to 0. In dry insulation, the resistance R_{abs} is high, the charge of the capacitor C_{abs} flows slowly, so the initial value of the current I_{abs} is small, and the current takes a long time to drop;

(3)   through conduction current I_{thr}, flowing through a resistance R_{thr}, depending on the characteristics of both external pollution of the insulation and on the paths of through leakage; this current is established almost instantly and does not change over time.

The total insulation resistance, measured by a megohmmeter, is inversely proportional to the sum of the indicated current components; at the beginning of measurement it has the lowest value, and then, as the current falls, I_{abs} increases, reaching a steady-state value determined by the current I_{thr}. In order to have comparable results, the insulation resistance is measured 60 s after the voltage is applied, although in some cases the current I_{abs} has not completely dropped by this time.

Measurement of the insulation resistance of transformer windings is carried out in accordance with GOST 3484.3–88. The insulation resistance value indicates the average condition of insulation and decreases as this condition worsens, mainly due to moisture and pollution. To estimate the condition of the transformer insulation, the insulation resistance of all windings is measured. When measuring, all the winding leads of the same voltage are connected together. The rest of the windings and the transformer tank must be grounded.

When assessing the condition of insulation, simultaneously with a measurement of the resistance R_{60} the absorption coefficient is measured. The absorption coefficient is the ratio of insulation resistance measured 60 s after the voltage is applied to the resistance measured after 15 s; its values do not depend on the geometric dimensions of insulation and characterize only the intensity of the drop in the absorption current. With the removal of moisture from the insulation, the absorption coefficient increases, and with moisture it decreases.

Table 2.6 Insulation resistance of transformer windings

Winding temperature, t °C	10	20	30	40	50
Insulation resistance 35 kV, Mohm	450	300	200	130	90
Insulation resistance 110 kV, Mohm	900	600	400	260	180
tgδ winding insulation,%	1,8	2,5	3,5	5,0	7,0

The value of absorption coefficient $K_{abs} = R_{60}/R_{15}$ must be no less than 1.3 at temperatures of 10 to 30 °C. For well-dried insulation, the values of absorption coefficient normally range from 1.3 to 2.0, see [34]. During commissioning and preventive tests, insulation resistance is measured in accordance with schemes used by the manufacturer, and additionally also in the insulation zones. During preventive tests, measurements are only allowed in the isolation zones.

The insulation temperature when measuring the insulation resistance of windings must be no less than

10 °C—for transformers with voltage up to 150 kV inclusive;
20 °C—for transformers with a voltage of 220–750 kV.

The insulation resistance of the windings of transformers up to 35 kV inclusive should be no lower than the values presented in Table 2.6.

Being combined with other control methods, this method can be useful for detecting contamination and moisture in insulation at relatively early stages. Measurements are usually carried out with a 2.5 kV megohmmeter using a clamp on the "shield" for measurement by zones. The insulation resistance is highly dependent on the winding temperature and the stability of the megohmmeter voltage. Thus, the difficulty of determining the temperature with reliability (by its resistance to direct current) reduces the value of insulation resistance as a parameter for estimating the winding condition.

The disadvantages of this method include restrictions on the temperature range at which the insulation resistance is measured, while the detection of damage occurs at the stage of insulation aging, which is close to critical. This method allows one to determine only gross defects in equipment: through burnout, strong moisture or contamination of insulation [7].

2.2.5 Measurement of Winding DC-Resistance

The measurement of winding resistance to direct current is made to examine the condition of electrical contact connections and the integrity of the electrical circuit of transformer windings. Defects identified during this measurement are caused by the following reasons:

– reakage of one or more of parallel wires in the taps;
– violation of soldering;

Fig. 2.25 Winding DC resistance measurement circuits

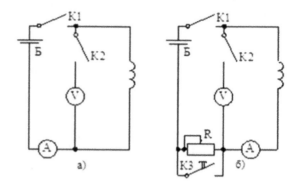

a)   б)

- poor-quality contact of connecting the winding taps to the bushings;
- poor-quality contact in the off-circuit tap-changers and on-load voltage regulation devices (OLVR) of the transformer;
- incorrect installation of an off-circuit tap-changer.

In installation conditions, resistance is usually measured by the ampermeter and voltmeter method. Figures 2.25a, b show two schematic diagrams of connecting instruments for measurements. The diagram in Fig. 2.25a is used when measuring small resistance values: starting from fractions of Ohm to several Ohms, and the circuit in Fig. 2.25b is used when measuring large values of resistance. A correct choice of measurement circuit eliminates significant errors due to voltage drops in the devices, which are usually not taken into account when calculating the resistance value.

Basically, in practice, the circuit shown in Fig. 2.25a is used. In this scheme, the current and voltage circuits are separated. They are made with separate wires in order to exclude from the measured resistance the resistance of the wires of the current circuits and the transition resistances at the points of connection of the voltage circuits to the inputs of the transformer. The voltage measuring circuit must be connected directly to the current-carrying pins of the bushings of the tested winding. Typically, resistance is measured at voltages up to 24 V and currents up to 10 A. In this case, the excess current should be no more than 20% of the rated winding current.

For single-phase transformers, the resulting values should not differ by more than 2% of the values specified in the passport, at the same temperature and with the same control taps.

For three-phase transformers, the resistances measured on the same branches of different phases should not differ from each other by more than 2%, unless there are special instructions in the passport.

The resulting values of the winding resistance to direct current lead to the temperature specified in the transformer passport. They are determined using the formula

$$R_x = \frac{R_0(235 + t_x)}{235 + t_0}, \tag{2.5}$$

where R_x is the resistance value at the temperature t_x specified in the passport, Ohm; R_0 is the resistance value at the measurement temperature t_0, Ohm; t_0 is the measurement temperature, $^\circ C$; t_x is the temperature $^\circ C$ specified in the passport.

The temperature of the upper layers of oil is taken as the temperature of an oil transformer which has not yet been turned on and heated, provided that the resistance is measured no earlier than 30 min later after oil filling for transformers up to 1000 kVA inclusive, and no earlier than 60 min for transformers of higher power.

The considered circuits for measuring the resistance of windings and the method itself have the following disadvantages:

– measurements must be made at temperatures no lower than $+10\ ^\circ C$;
– all devices used must be of high accuracy class;
– only highly developed defects are registered.

2.2.6 Short-Circuit Resistance Monitoring

This control method, due to its relative simplicity, is widely used in power systems, both when testing transformers for short-circuit resistance, and in operation.

The essence of the method is exemplified in Fig. 2.26. It consists in measuring the voltage drop across the resistances in each phase of a transformer and the current flowing through them, and in determining the resistance value from the result obtained.

Let us denote by R_1 and L_{S1} the active resistance and leakage inductance of the primary winding of the transformer; R_2 and L_{S2} are the values of active resistance and scattering inductance of its secondary winding reduced to the primary winding; R_0 and L_0 are the active resistance and inductance of the primary winding, which

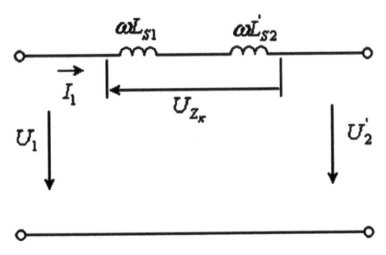

Fig. 2.26 Equivalent circuit of a transformer

determine the no-load current of the transformer. For power transformers, the typical relation between element parameters in a circuit meets the conditions

$$\omega L_{S1} >> R_1, \ \omega L'_{S2} >> R'_2 \ \omega L_{S1} << \omega L_0 >> \omega L'_{S2}$$

Under these assumptions, the electrical circuit of a transformer can be presented in a simplified form, Fig. 2.26.

By measuring the voltage U_{Z_K} and current I_1, one can calculate the modulus of Z_K, since the leakage inductance of a winding is directly related to its geometric dimensions; changes in the latter are unambiguously reflected in changes in the values of $L_{S1} + L_{S2}$. Therefore, monitoring changes in leakage inductances is a direct method for measuring winding deformation. It should be noted that the leakage inductance of transformer windings is small, and it is necessary to measure its change with high accuracy. Therefore, a measurement should be organized according to the first harmonic, taking measures to minimize the level of higher harmonics, first of all, the third harmonic, which occurs when the magnetization reversal of a transformer core is due to the nonlinearity of the magnetization curve of a steel core. This creates additional inconveniences in control. The main disadvantage of the method is that it is highly sensitive only to those types of deformations of transformer windings that lead to a change in the size (volume) of the scattering channel. Therefore, having a relatively good sensitivity to the loss of radial stability of windings, due to the fact that this increases the volume of the main scattering channel, the Z_K method turns out to be insensitive to other types of damage, such as lodging of turns, slipping of winding turns, and unwinding.

References

1. Lvov MYu, Kutler PP (2003) Physicochemical methods in the practice of estimating the condition of power transformers under operating conditions: study guide. IUE GUU, VIPK-energo, Moscow, 20 p (in Russian)
2. Müller R, Soldner K, Schliesing H (1977) Elektrizitätswirtschaft 76(11):345–349
3. Dörnenburg E, Hutzel O (1977) ETZ-A 98(3):211–215
4. Rogers RR (1978) IEEE and IEC codes to interpret incipient faults in transformers using gas in oil analysis. IEEE Trans Electr Insul 13(5):348–354
5. IEEE Std C57.104-1991/IEEE Guide for the Interpretation of Gases Generated in Oil-immersed Transformers
6. Oommen TV, Ronnau RA, Girgis RS (19998) New mechanism of moderate hydrogen gas generation in oil-filled transformers. (United States). CIGRE Paris 1998, paper 12–206
7. Alekseev BA (2002) Condition monitoring (diagnostics) of large power transformers. Moscow: Publishing House of NTs ENAS, 2002. 216 p. ill. Basic electrical equipment in power systems: an overview of domestic and foreign experience (in Russian)
8. Darian LA (2008) Scientific basics of physical and chemical diagnostics of high-voltage oil-filled electrical equipment with capacitor-type insulation: dissertation for the degree of Doctor of Technical Sciences, Novosib. State Tech. Univ. – Novosibirsk, 445 p (in Russian)
9. Darian LA, Arakelyan VG (1998) Problem and achievement a physico-chemical diagnostics of oil-filled electrotechnical equipment. Darian LA, Arakelyan VG. CWIEME. Berlin

10. Moflmann A (1999) Pahlavanpour B. New guidelines for interpretation of dissolved gas analysis in oil-filled transformers. ELECTRA, No. 186, October, 1999

11. Guidelines for the diagnosis of incipient failures in transformer equipment based on the results of chromatographic analysis of gases dissolved in oil. RD 153-34.0-46.302-00. Moscow, 2001 (in Russian)

12. Condition monitoring and maintenance of electrical machines (1989) Proceedings Coll SC11 CIGRE and CEA, Montreal, Quebec

13. Dyagteryaev SA, Dolin AP (2003) Basic concepts for complex diagnostic tests of power transformers. Electro (2) (in Russian)

14. Kasatkina TE, Lvov MYu (2001) Experience of using chromatographic analysis of gases for estimating power transformer conditions. New in the Russian electric power industry, 7. (in Russian)

15. Lvov MYu (2002) Relevance assessment of control indicators for technical conditions of transformer equipment insulation. Electr Stat (12). (in Russian)

16. Diagnostic guidelines for incipient failures based on chromatographic analysis results for gases dissolved in transformer oil. RD 34.45–51.300-97. – Moscow, 1989 (in Russian)

17. Lvov MYu, Chichinsky MI, Lvov YuN et al (2002) Standardization of indicators for estimating the insulation wear of power transformer windings. Electr Stat (7). (in Russian)

18. Osotov VN (1995) Research on diagnostic possibilities for pressing forces of transformer windings based on vibration characteristics. Osotov VN, Petrishchev LS, Saltanov VM. Electr Stat 8:32–37. (in Russian)

19. Gervits MN (1997) Diagnostic methods for pressing forces of transformer windings. Gervits MN, Osotov VN, Petrishchev LS. Electr Stat 5:58–60. (in Russian)

20. Rusov VA (1998) Pressing control for windings and magnetic conductors of large transformers based on vibration parameters. Rusov VA. Electr Stat 6:52–57. (in Russian)

21. Emelyanov VI (200) Experience in methodology application for determining residual pressing forces of power transformer windings. Emelyanov VI, Ulyanov AM, Ruschinsky VN, Amromin AL. Abstracts of the X International Scientific and Technical Conference "Transformer Engineering-2000". September 19–21. Zaporizhzhia. – PO ZTV, pp 145–149. (in Russian)

22. Zhou H (2016) Transformer winding fault detection by vibration analysis methods. Zhou H, Hong K, Huang H, Zhou J. Appl Acoust 114:136–146

23. Bartoletti C (2004) Vibroacoustic techniques to diagnose power transformers. Bartoletti C, Desiderio M, Carlo D, Fazio G, Muzi F, Sacerdoti G. Appl Acoust 19:221–229

24. Garcia B (2006) Transformer tank vibration modeling as a method of detecting winding deformations. Garcia B, Burgos S, Alonso A. Part I: theoretical foundation. IEEE Trans Power Deliv 21:157–163

25. Garcia B (2006) Transformer tank vibration modeling as a method of detecting winding deformations. Garcia B, Burgos S, Alonso A. Part II: experimental verification. IEEE Trans Power Deliv 21:164–169

26. Tettex Instruments AG, Partial Discharge Measuring Technique, Tettex Information, No. 21, 1994. http://www.maceyselectrical.com.au/Haefely_Hipotronics_DDX9101_Partial_Discharge_Detector.html

27. https://www.pfiffner-group.com/products-solutions/details/ddx9121

28. Sikorski W, Siodla K, Moranda H, Ziomek W (2012) Location of partial discharge sources in power transformers based on advanced auscultatory technique. IEEE Trans Dielectr Electr Insul 19(6):1948–1956

29. Carlson A, Fuhr J, Schemel G, Wegscheider F (2010) Testing of power transformers and shunt reactors, ABB Ltd. Transformers, Zurich, Switzerland, ISBN 3-00-010400-3, 2nd Edition, 2010

30. Rusov VA (2011) Measurement of partial discharges in high-voltage equipment insulation. Yekaterinburg: UrGUPS, 2011, p 368 – ISBN 978-5-94614-177-2. (in Russian)

31. Jia Y, He P, Huo L (2020) Wireless sensor network monitoring algorithm for partial discharge in smart grid. Electr Power Syst Res 189:1–12. https://doi.org/10.1016/j.epsr.2020.106592

32. Braeunlich R, Haessig M, Fuhr J, Aschwanden Th (2000) Assessment of insulation condition of large power transformer by on-site electrical diagnostic methods. IEEE International symposium on electrical insulation, Anaheim CA, IEEE Publication 00CH37075, 2000, p 368
33. Fuhr J (2008) Condition assessment of insulating systems of generators. Monitoring and diagnosis meeting, arnhem
34. "Electric Grids" Website / Start / Transformers / Practice / Method of measuring insulation resistance and absorption ratios of power transformers

Chapter 3
Diagnostics of Transformer Windings Condition by Probing Pulses of Microsecond Duration

3.1 Physical Basis and Development of Pulsed Diagnostics

Low-voltage pulsed diagnostics (LVPD) was proposed in 1966 at the Institute of Electrical Engineering (Warsaw, Poland) [1] and was instantly recognized by the world's leading centers for the development and implementation of new control technologies for transformer condition [2]. The measurement scheme originally used to implement this technique is shown in Fig. 3.1.

The method application is illustrated by the example of a 50 kVA, 15/0.4 kV transformer. Figure 3.2 shows the active part, and Fig. 3.3 shows the voltage curves recorded at the movement by 6 mm of individual coils in different locations of phase A in the 15 kV winding. The pulses were fed to the 15 kV winding of phase A; the 0.4 kV winding is short-circuited.

A standard storm impulse with parameters 1.2/50 µs was used as a probe pulse. It was suggested that changes in the curves being compared indicate mechanical displacements in the windings or short-circuited turns.

By the nature of changes in the curves, it is possible to draw some conclusions regarding the type of damages and their gradual amplification during next short circuits, as well as regarding the site of their occurrence in the winding. Changes in the voltage curves are caused by changes in capacitance and inductance (mainly longitudinal) at the site in the winding where short-circuit or mechanical movement of the turns have occurred. These changes are reflected in the shape of the curves. Comparison of the current and previous measurements allows one to make a well-grounded conclusion about the winding condition.

Based on the results of the original experiments described in [1, 2], the following conclusions have been drawn.

1. Fault detection methods based on measuring the effects caused by changes in the transversal magnetic scattering flux can be quite sensitive, but they are also very laborious.

2. Measurements of the higher harmonic content in the idling current do not allow detecting small displacements of windings, and, at the same time, require a

V. Ya. Ushakov et al., *Transformer Condition Control*, Power Systems, https://doi.org/10.1007/978-3-030-83198-1_3

Fig. 3.1 Measurement scheme first used to implement the low-voltage pulsed method. R is a measuring resistance; W is a broadband amplifier; KO is a cathode oscilloscope

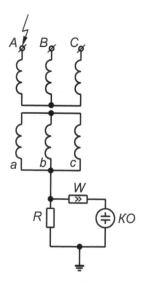

Fig. 3.2 Active part of a 50 kVA, 15/0.4 kV transformer, the fifth coil of the right phase from the top being moved 6 mm down

supply of voltage that is stable in both the magnitude and the sinusoidal shape. The measurement results allow one to identify the phase that has been damaged, but do not indicate the site of damage in this phase.

3. The above method of low-voltage pulses does not have the mentioned drawbacks and is easy to use. Measurements can be carried out after each SC, and

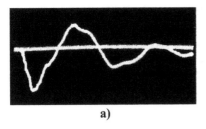 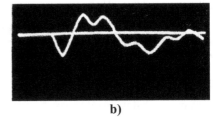

a)   b)

Fig. 3.3 Change in the shape of low-voltage probe pulses with intentional downward movement of two upper coils of phase A: **a** before moving the coils, **b** after moving the coils

the occurrence and growth of displacements in the windings can be observed. This permits a dynamic strength test to be terminated even before the transformer has been destroyed. When using this method, there is no need to open a transformer. The sensitivity of the method is very high. It allows detecting mechanical movements of even individual turns, and, with some experience, gaining valuable information about the size and type of damage.

4.   The LVP (low-voltage pulsed) method can also be used in transformer operation. Making records of voltage oscillograms before putting a transformer into operation and then during periodic control examinations and comparing them with each other provide valuable data on the mechanical condition of the windings.

The drawbacks of this method include the fact that high reproducibility of measurement results is reachable only when ensuring complete identity of measurements, the time between which can span years (scheme and procedure of measurements, cables and connectors used, their relative position during tests, etc.). In addition, normograms obtained on a new or known-good transformer are required. Interpretation of measurement results calls for highly qualified operating personnel. As shown by further practice, the use of a storm impulse as a probe is ineffective. The inefficiency is explained by the relatively low content of high-frequency components in the integral spectrum of a storm impulse. Due to this physical limitation at the initial stage of development, the LVP method did not succeed in detecting incipient faults. As further research has demonstrated, it is an expansion of the high-frequency range of a probing pulse that has a decisive effect on the sensitivity of a diagnostic procedure. Therefore, a gradual shift of the pulse shape from storm to rectangular was a natural stage in the development of the LVP method and was brilliantly embodied in nanosecond probing.

3.2   Development of Pulsed Diagnostics Technology

Substantial experience of using this method was accumulated in the USSR, where its development and application began in 1973 at the Laboratory of transformer diagnostics of the All-Union Electrotechnical Institute, presently, the All-Russian

Electrotechnical Institute, (FGUP VEI), see [3]. The method was improved and given the name of ***pulsed defectography***. Improvement was carried out in the direction of standardizing the probe pulse, namely, its shape and duration, as well as in the development of a system for recording response signals based on electronic circuits. Attention was paid to improving the accuracy of diagnostic results, in particular, the parameter "pair correlation ratio" was suggested as a criterion for comparing the normogram and the defectogram.

The essence of the method is as follows. A voltage pulse with an amplitude of 100–500 V is applied to one of the terminals of a transformer, which does not pose a danger to the insulation of the transformer windings. Then, the response signals of the windings of the other phases are recorded and analyzed [3]. The response signals obtained during preventive tests are compared with the signals obtained when probing the windings of a known-good transformer. Their differences are due to the fact that even with insignificant mechanical movements in the windings, the capacitance of individual elements (inter-turn and inter-coil capacitance, capacitance for a tank or magnetic conductor) can change significantly, and with significant deformations, the inductance of deformed elements can also change. This leads to a change in the natural frequencies of oscillations in individual circuits, which is reflected in the oscillograms of pulsed currents and voltages. Monitoring the mechanical condition of windings using the LVP method has a higher sensitivity than measuring the short-circuit resistance. One of the first modifications of devices of the series "Impulse", developed and successfully tested at the Laboratory of Impulse Defectography at FGUP VEI, is shown in Fig. 3.4. By 1978, more than 30 transformers had been tested using the Impulse-2 installation, after which it was considered expedient to further develop this direction of diagnostics [3].

All major stands for testing windings for SC resistance were equipped with installations of the Impulse type. Subsequently, the installations were upgraded for their mobile use in a number of electrical power grids and at large enterprises with a well-developed transformer stock. The use of the Impulse diagnostic units made it possible to detect mechanical deformations of transformer windings with a rated voltage of 35–500 kV and a capacity of 6300–500000 kVA. This made it possible to take transformers out of service in a timely and reasonable manner, while preventing possible serious accidents. Disassembly of transformers confirmed the data on the presence of winding displacements [4]. Since 1983, test benches have been equipped with the Impulse-5 measuring system, the appearance of which is shown in Fig. 3.5.

With the progress of microprocessor technology, the measuring complex was modified [5]. Figure 3.6 shows the appearance of the Impulse-7 measuring system based on microprocessor technology.

Later on, the system went through several stages of modification in the direction of reducing the size while increasing the efficiency of the condition monitoring process. Figure 3.7 shows a general view of the latest version of the measuring system, which includes a personal computer with a built-in signal processing program.

The measurement technique consists in feeding a rectangular probing pulse, with a voltage of 100–500 V and a duration of 1 μs, from a special generator to one of the windings or to the neutral terminal of a transformer. At the same time, oscillography

Fig. 3.4 Impulse-2
installation

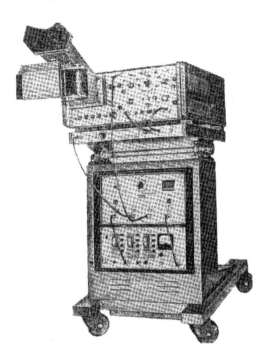

of the response of the windings to the effect of the applied pulse takes place: the oscillogram of voltage across the measuring resistances which are connected to the other windings is recorded, as shown in Fig. 3.8. The method is based on sequential impulse defectography [6]. First, normograms are recorded from a known healthy transformer at a manufacturing plant or at a power plant before it has been put into operation. They are subsequently compared with defectograms, being oscillograms obtained during subsequent measurements after the transformer has been put into operation. The condition of transformer windings is determined by comparing normograms and defectograms between themselves according to a certain method. Deviations of the defectogram from the normogram indicate the presence of electrical damage or mechanical displacement of the windings. If there are no normograms available, then the analysis of winding condition is carried out by comparing the oscillograms of different phases of the transformer.

The high efficiency of the LVP method in detecting residual deformations of power transformer windings is provided by a high sensitivity of the method to changes in geometrical dimensions. Even small mechanical movements in the windings significantly change the capacitance of their individual elements (turn-to-turn, inter-coil), and significant deformations lead to changes in the inductance of the deformed elements. All of the above changes the natural frequencies of oscillations, which is recorded in oscillograms of current and voltage pulses [7]. Figure 3.9 shows a diagram of the pulse defectography of a transformer, and Fig. 3.10 shows an equivalent scheme of one of its windings.

Fig. 3.5 Impulse-5 system
for diagnostics of the
mechanical condition of
power transformer

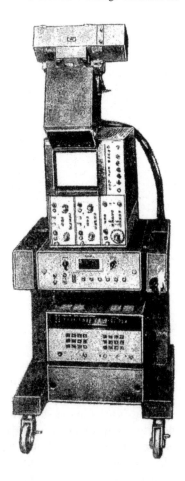

Despite the relatively high (in comparison with other techniques) sensitivity of
the LVP method and its long-term improvement (in the period from 1966–1988),
its error remained unacceptably large in a number of cases of practical application.
This was due to inaccuracy and subjectivity in the analysis of defectography results.
Figure 3.11 shows oscillograms of defectrography, which imply that the identification
of a damage type is subjective and requires highly qualified personnel [1, 2, 8].

Despite the demonstrated technological advantages of pulsed defectography and
its recognition as the most accurate and perfect technique (as of the end of the
1980 s), it did not continue and stopped developing in the early 1990 s for a number
of technical, organizational and economical reasons.

Researchers at FGUP VEI have shown experimentally that mechanical damage to
transformer windings due to electrodynamic effects can be accompanied by essen-
tially smaller changes in Z_{SC}, which has been noted in [5, 6]. Some of the research
results are shown in Fig. 3.12.

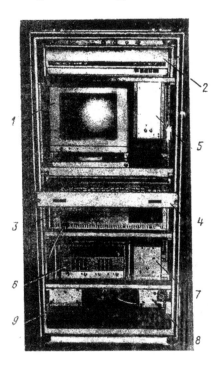

Fig. 3.6 System for diagnostics of the mechanical condition of power transformer windings, Impulse-7, based on microprocessors: 1 is an oscilloscope, 2 is a printing device, 3 is a computer system unit, 4 is a keyboard, 5 are cable connctions to a tested object, 6 is a pulse generator, 7 is an- amplifier, 8 is a fan, 9 is a shockproof shell

Fig. 3.7 Impulse-10 system for diagnostics of the mechanical condition of power transformer windings

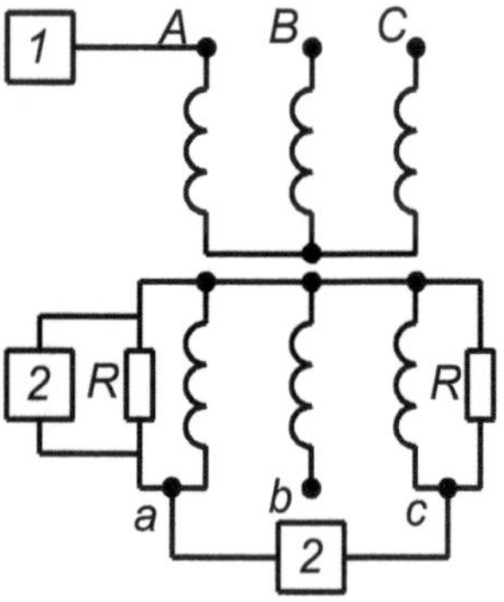

Fig. 3.8 Measurement scheme when implementing the LVP method: 1 is a pulse generator; 2 is an oscilloscope; R is a measuring resistor

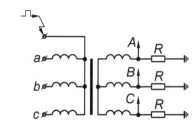

Fig. 3.9 Scheme of pulse defectography of a transformer

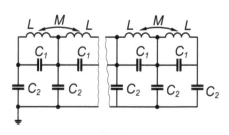

Fig. 3.10 Equivalent circuit of a transformer winding

The main advantages of the LVP method are its high sensitivity to a large number of damages, the ability to determine not only the phase, but also the winding in which the damage has occurred. Основные преимущества метода НВИ состоят в высокой чувствительности к большому количеству повреждений, в возможности определять не только фазу, но и обмотку, в которой произошло повреждение.

However, the method does not allow a quantitative assessment of the resulting residual deformations, requires disconnecting and unbussing a transformer, which makes it impossible to use monitoring systems under operating voltage that control the mechanical condition of windings at the flow of through SC currentsDisplacements arising from the flow of through SC currents through transformer windings do not always lead directly to turn-to-turn short circuits, i.e., they do not always cause instant operation of transformer protection. Untimely withdrawal of a transformer

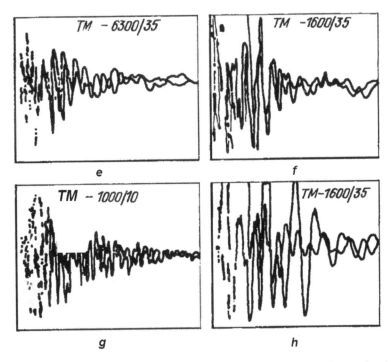

Fig. 3.11 Typical oscillograms for the most common type of damage to windings under electro-dynamic effects in TM (High Voltage Oil-filled Transformer) transformers for a voltage class up to 35 kV: **a** no damage; **b** unpressing the windings; **c** loss of pressing, mutual movement of the windings; **d** loss of axial stability; **e, f** loss of radial stability of the inner winding; **g, h** electrical damage

with deformed windings for repair leads, as a rule, to further displacements and turn-to-turn short circuits even without short-circuits and a forced decommissioning of the transformer. Timely detection of displacements makes it possible to take a transformer out for repair with subsequent replacement of damaged units and to make the most of undamaged ones [5–8].

3.3 Frequency Analysis as a Development of Pulsed Diagnostics

The method was proposed and implemented in Canada in 1978 by a research laboratory at Ontario-Hydro. *The FRA (Frequency Response Analysis)* method is based on comparing a currently obtained frequency response of a transformer winding with characteristics previously taken using the same transformer, or a transformer of the same type, before it has been put into operation. In practice, the most common measurement is that of reactive conductivity in the frequency range up to 2 MHz. The

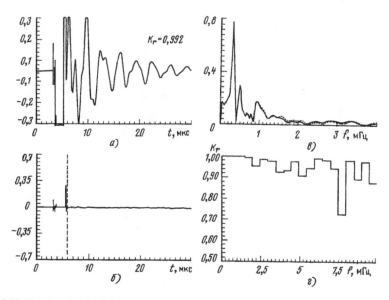

Fig. 3.12 Results of pulsed defectography of a transformer: TM-630/35: **a** normogram (before testing) and defectogram (after testing) with overlapping in the absence of changes; **b** oscillogram of their differences; **c** frequency spectra of the normogram and defectogram; **d** pair correlation function showing in what frequency range the changes have occurred; K is the pair correlation ratio of normogram and defectogram

measurement procedure using the frequency analysis method involves recording the applied pulse and the corresponding response using high-precision analog-to-digital converters (ADC). The results are then converted to the frequency domain using the Fast Fourier Transform (FFT) algorithm. The transfer function calculated in this way will depend only on the parameters of the test object and will not depend on the applied signal and the measurement scheme. Changes in the object can be clearly identified and separated from the manifestations of external factors, which greatly facilitates the analysis of the results [9].

In parallel to this method, another one was developed: instead of a pulsed action, a sinusoidal voltage was applied to a winding input, which varied in frequency over a wide range, and the response voltage was recorded from the other inputs. Based on the results, an amplitude-frequency characteristic was built, being the response of the windings to the impact. Both methods are widely used. Figure 3.13 shows a diagram of taking the amplitude-frequency characteristics of an inspected transformer using the frequency analysis method.

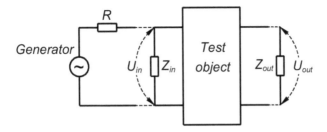

Fig. 3.13 Schematic diagram of frequency analysis measurements

The response of windings is analyzed by the following manifestations:

- difference between the response signals of individual phases of a transformer;
- difference between response signals of transformers of similar design;
- alteration in the amplitude-frequency characteristics obtained on one transformer, but at different times.

A two-channel analog-to-digital converter records a signal through channel No. 1, which is fed to a winding input, and through channel No. 2, the response of the winding to the effect of the signal (probe pulse). After that, the transfer function is calculated, which is the ratio of the spectra of the input and output signals. The degree of difference between the transfer functions, calculated before and after exposure of the transformer to electrodynamic short-circuit forces or other mechanical influences, is carried out using different analysis tools.

Figure 3.14 shows an example of frequency response and phase-frequency response (PFR) of the higher voltage windings of a 25 MVA transformer with a measurement range of $0 \div 2$ MHz, recorded at different times.

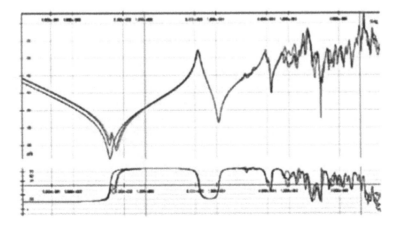

Fig. 3.14 Amplitude-frequency and phase-frequency characteristics of the higher voltage windings of a 25 MVA transformer

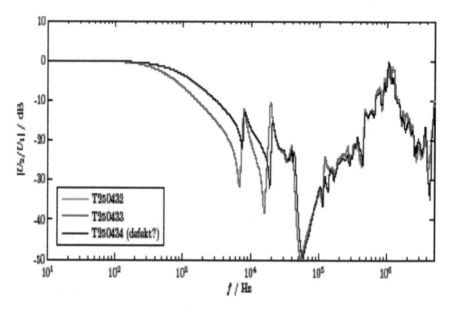

Fig. 3.15 Frequency characteristics of a single-phase 110/10 kV, 40 MVA transformer; curve T 250434 corresponds to a defective winding

The type of frequency response and phase frequency response, shown in Fig. 3.14, indicates a change in the mechanical condition of transformer windings. When analyzing the results of measurements of the frequency spectrum, the influence of various parameters is taken into account, such as: the grounding of cables, especially manifested at high frequencies, the presence or absence of oil in the transformer during its examination.

Figure 3.15 shows the frequency response of a 50 MVA transformer in the frequency range from 0 to 1 MHz with and without oil in the transformer tank. It can be seen that there is a shift in the characteristic over the entire range due to a change in the values of the corresponding values of capacitance.

Figure 3.15 shows the amplitude-frequency characteristics of a transformer operating in a "generator-transformer" unit. The transformer windings are star-connected. In order to observe a defect, experience is required, since even when using highly sensitive and very high-quality measuring equipment (for example, that of the Swiss company *Haefely Test AG*) there is a high probability of missing a defect. This is one of the reasons for the lack of sensitivity of this diagnostic method. It should be noted that the frequency response shown in Fig. 3.16 exemplifies the initial stage of a winding defect development. As shown by the subsequent opening of the transformer and a thorough visual inspection of the winding, the defect was due to poor-quality pressing. The damage had possibly occurred during transportation of the transformer. Subsequent disassembly of the transformer fully confirmed the presence of a winding defect caused by short-circuit currents.

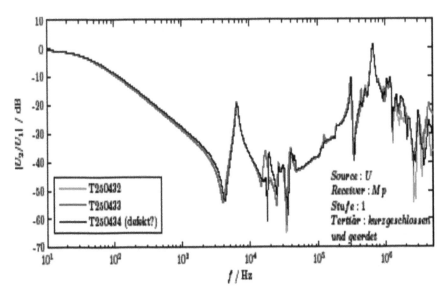

Fig. 3.16 Frequency response of a 273 MVA transformer in a generator-transformer unit with a star type of winding connection

The nature and degree of development of an identified damage do not raise doubts that an emergency mode is inevitable if such a transformer continues to operate. Consequently, the malfunction has been detected in due time: at the initial phase. Obviously, at the first exposure to short-circuit currents, the transformer would fail. The characteristic "overshoot"/"burst" at the bottom right, observed in Fig. 3.16, indicates the presence of a defect in the winding. The result of analysis according to the given frequency characteristics is fully confirmed by opening the transformer, which is recorded in the photograph shown in Fig. 3.17.

When carrying out such measurements, it is necessary to carefully and correctly ground when connecting the measuring system in order to avoid interference and distortions that penetrate the measuring circuits. During measurements, the grounding of the transformer housing and the recording equipment should be carried out at the same point. Measurements taken at different times with grounding at different points distort the frequency response, which leads to an incorrect assessment of the winding condition. The cost of neglecting such, at first glance, secondary factors can lead to an incorrect diagnosis and, as a consequence, subsequent unjustified expenses for the revision of the transformer. The popularity of the *FRA* method is growing, and a large number of studies are devoted to it in the world's leading high-voltage laboratories [10–20]. *FRA* technology is discussed at CIGRE colloquia and IEC meetings on the formation of test and measurement standards [21, 22]. Despite the absolute advantages of the *FRA* method, the percentage of errors remains quite high. In a number of cases, in the presence of a defect, the diagnosis was "healthy", i.e., the defect was not recognized. At the same time, there were situations in which

Fig. 3.17 Photograph of a defective winding corresponding to the frequency characteristics shown in Fig. 3.16

a transformer with serviceable windings was rejected and taken out for repair. The materials of Chaps. 2 and 3 allow one to conclude that there is no reliable, proved and effective method for monitoring the condition of transformer windings, despite the variety of existing methods and the long history of their development. To create a new technology for monitoring the condition of active parts of power transformers that should fully meet the requirements of the electric power industry of the 21st century, the efforts of developers should be focused on increasing the sensitivity, accuracy and reliability of the method for detecting winding defects at an early stage of their development. As will be shown in Chap. 4, monitoring the condition of active parts can be performed without using normograms, as well as without removing the operating voltage of a high-voltage network and disconnecting the transformer.

References

1. Lekh V, Tyminskiy L (1966) Novyi metod indikaschii povrezhdenii pri ispytanii trasforma-torov na elektricheskuyu prochnost' [New method of fail diagnostics at the dynamic reliability transformer test]. Elektrichestvo—Electricity 1(1):77–81. In Rus
2. Rogers EI, Gillies DA, Humbard LE (1972) Instrumentation techniques for low voltage impulse testing of transformers. IEEE, PAS 91(3)
3. Avetikov GV, Levitskaya EI, Popov EA (1978) Impulsnoe defektographirovanie transforma-torov pri ispytaniyakh na elektrodinamicheskuyu stoikost' [Impulsed defectography of trans-formers at the electrodynamic reliability test]. Elektrotechnika – Electrotechnic (4):53–57. In Rus

4. Konov YuS, Korolenko VV, Fedorova VP (1980) Obnaruzhenie povrezhdenii transformatorov pri korotkikh zamykaniyakh [Fail detection of transformers due to short circuit]. Elektricheskie Stanschii – Electric Power Stations (7):46–48. In Rus

5. Alikin SV, Drobyscevskiy AA, Levitschaya EI, Filatova MA (1990) Kolichesvennaya oshcenka rezultatov impulsnogo defectografirovaniya obmotok silovykh transformatorov [Quantitative assessment of the of pulse defectography results of power transformer windings]. Elektrotechnika – Electrotechnic (5):75–76. In Rus

6. Alikin SV, Drobyscevskiy AA, Levitschaya EI, Filatova MA (1991) Diagnostika obmotok silovykh transformatorov metodom nizkovoltnykh impulsov [Power transformer winding diagnostics by low-voltage pulsed method]. Elektrotechnika – Electrotechnic (12):30–35. In Rus

7. Khrennikov AYu, Kikov OM (2003) Diagnostika silovykh transformatorov v Samaroenergo metodom nizkovoltnykh impulsov [Power transformer diagnostics by low-voltage pulsed method in Samaropenergo]. Elektricheskie Stanschii – Electric Power Stations (11):49–51. In Rus

8. Khrennikov AYu (2003) Opyt obnaruzheniya ostatochnykh deformaschiy obmotok silovykh trasnformatorov [Experience of mechanical deformation detection of power transformer windings]. Energetic – Energetic (7):18–20. In Rus

9. Malewski R (1995) Monitoring of winding displacements in HV transformers in service. Malewski R, Yu A. Khrennikov, Shlegel OA, Dolgopolov AG. A.G. CIGRE Working Group 33.03. Italy. Padua

10. Guillen D, Olivares-Galvan J, Escarela-Perez R, Granados-Lieberman D, Barocio E (2019) Diagnosis of interturn faults of single-distribution transformers under controlled conditions during energization. Measurement 141:24–36

11. Zhaoa X, Yaoa C, Abu-Siadab A, Liaoa R (2019) High frequency electric circuit modeling for transformer frequency response analysis studies. Electr Power Energy Syst 111:351–368

12. Cheng Q, Zhao Z, Tang C, Qian G, Islam S (2019) Diagnostic of transformer winding deformation fault types using continuous wavelet transform of pulse response. Measurement 140:197–206

13. Velasquez R, Lara J, Melgar A (2019) Converting data into knowledge for preventing failures in power transformers. Eng Failure Anal 101:215–229

14. Zarkovi M, Stojkovi Z (2017) Analysis of artificial intelligence expert systems for powertransformer condition monitoring and diagnostics. Electr Power Syst Res 149:125–136

15. Larin VS, Matveev DA, Volkov AYu (2020) Interpretatsiya chastotnykh kharakteristik silovykh transformatorov s primeneniyem matritsy provodimostey obmotok [Interpretation of the frequency characteristics of power transformers using the matrix of the conductivity of the windings]. Elektrichestvo – Electricity (5):13–22. In Rus. https://doi.org/10.24160/0013-5380-2020-5-19-25

16. Larin VS (2018) Analiz chastotnykh kharakteristik dlya lokalizatsii korotkikh zamykaniy v obmotkakh transformatorov [Analysis of frequency characteristics for localization of short circuits in transformer windings]. V.S. Larin. Elektrichestvo – Electricity (4):14–25. In Rus

17. Larin VS (2015) Mirovyye tendentsii razvitiya transformatornogo oborudovaniya [World trends in the development of transformer equipment]. V.S. Larin. Elektrichestvo – Electricity 8:20–26. In Rus

18. Tenbohlen S (2016) Diagnostic measurements for power transformers review. Tenbohlen S, Coenen S, Djamali M, Andreas Muller A, Samim M, Siegel M. Energies 9(347), 25 p. https://doi.org/10.3390/en9050347

19. Wang M (2005) Improved detection of power transformer winding movement by extending the FRA high frequency range Wang M, Vandermaar A, Srivastava K. IEEE Trans Power Delivery 20(3):1930–1938

20. Zhao Z (2016) Determination of nanosecond pulse parameters on transfer function measurement power transformer winding deformation. Zhao Z, Yao C, Hashemnia N, Islam S. IEEE Trans Dielectr Electr Insul 23(6):3761–3770

21. CIGRE Working Group A2.26 (2008) Mechanical–Condition Assessment of Transformer Windings Using Frequency Response Analysis (FRA). Brochure 342
22. Measurement of frequency response (2012) IEC 60076–18

Chapter 4
Probing Transformer Windings with Nanosecond Pulses

This chapter presents the technologies for transformer windings condition controlled by means of nanosecond pulses. Main steps of development of physical as well as mathematical models are shown. The stages of improving the pulsed method based on a short probing pulse of the nanosecond range, the results of experiments on identifying the radial and axial displacements of the winding, studies of the effect of the duration and shape of the probing pulse on the sensitivity of the diagnostic procedure, and the stages of developing a mathematical as well as physical model of a power transformer are consistently presented. It is shown that short probing impulse diagnostics technology has potential advantage for application in electrical energy systems. Possible prospect ways of improving the technology for next level are given.

4.1 Physical Prerequisites for Increasing the Efficiency of PD Technology by Reducing the Duration of Probe Pulses

As established at the initial stages of mastering the technology of pulsed defectography, the efficiency of the LVP method for detecting residual deformations of power transformer windings is ensured by its high sensitivity to changes in the geometric dimensions of an object under study. Even the smallest changes in the condition of winding turns change the natural vibration frequencies, which is recorded in the oscillograms of pulse currents and voltages and makes it possible to establish the presence of a defective winding condition.

At the same time, the LVP and FRA methods do not always reflect the actual condition of a winding (presence or absence of defects) and have a relatively low sensitivity due to an insufficient saturation of the probing signal spectrum with the high-frequency component as part of the probing pulse (in the LVP method) and the frequency signal (in the FRA method). The standard storm impulse and impulses

V. Ya. Ushakov et al., *Transformer Condition Control*, Power Systems,
https://doi.org/10.1007/978-3-030-83198-1_4

with other parameters used in the LVP method have an upper frequency of the order of 500 kHz. The upper limit of the probing signal frequency range in the FRA method is 2 MHz.

The authors of the present monograph have proposed to carry out pulsed defectography with a pulse having a frequency range up to 20 MHz and higher, see [1–5]. This is implemented by feeding to one of the transformer windings a rectangular probe pulse having a duration of 500–50 ns, a rise time (front duration) of 20–5 ns and an amplitude of 100–200 V. The use of a probe pulse with such parameters allows recording a response which is formed only in the capacitive elements of a system. This makes it possible to identify the problem at the earliest stage of a defect condition development, since the transient currents in the inductive elements do not have time to form at such a short duration of probe pulses.

4.2 Elaboration of ID Technology on Physical and Mathematical Models of a Power Transformer

4.2.1 Creating a Physical Model of a Power Transformer

In physical modeling, a model is known to recreate the same principles of operation as those in a natural object, albeit modified in absolute values in accordance with the scale of modeling. One of the main advantages of physical modeling over mathematical modeling is the ability to carry out direct observations of the processes and phenomena under study. In this regard, for the development and study of a new method for monitoring the condition of transformer windings, a physical model of a high-voltage power three-phase two-winding transformer has been created and elaborated. The appearance of the model is shown in Fig. 4.1.

The low voltage (LV) winding is made of a 2 × 8 mm copper bus wound on a vinyl plastic cylinder and contains 20 turns. The outer diameter of the winding is 102 mm, the inner diameter is 86 mm, and the length is 370 mm. The LV winding is internal. The high voltage (HV) winding is made of copper wire with a diameter of 2 mm, wound on a PVC cylinder with a pitch of 4 mm, and contains 120 turns. The outer diameter of the winding is 160 mm, the inner diameter is 140 mm, and the length is 370 mm. The HV winding is external to the LV winding

Each coil contains taps and leads to simulate various defects and connecting measuring instruments. The three-phase two-winding structure is mounted on metal rods that simulate the core of a transformer. During our experiments, the physical model was connected to a Nanotest-1 type probe pulse generator and measuring devices, Tektronix type electronic oscilloscopes. Figures 4.2, 4.3 and 4.4 show photographs of the main elements of our physical model for a power transformer.

The Nanotest-1 generator allowed us to generate pulses with a duration of about 450 ns and a rise time of 15–20 ns having an amplitude adjustable from 50–500 V.

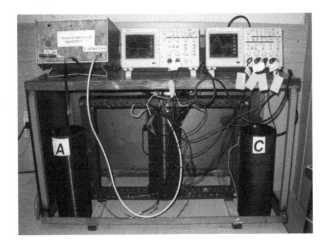

Fig. 4.1 General appearance of our physical model for a power transformer

Fig. 4.2 Physical model with LV windings

4.2.2 Detection of Turn-to-Turn Short Circuits

Defects caused by damage to turn insulation, such as turn-to-turn short circuits, make up a significant proportion of failures, whose further development leads the transformer to an emergency condition. Below we present the results of our experiments, during which a short circuit of two adjacent turns was created in the LV winding. The first stage of the experiments was made using a single-phase physical model, the diagram of which is shown in Fig. 4.5. A probe pulse with duration of 400 ns and amplitude of 200 V, whose oscillogram is shown in Fig. 4.6, was fed to the

Fig. 4.3 Physical model with HV windings

Fig. 4.4 LV winding of our
physical model connected to
a generator and an
oscilloscope

HV winding, and the response signal was taken from the LV winding. Response
oscillograms in the absence of a turn-to-turn closure of two adjacent turns a1 and
a2 (Fig. 4.5) are shown in Fig. 4.7a. Response oscillograms in the presence of a
turn-to-turn closure of two adjacent turns a1 and a2 (Fig. 4.5) are shown in Fig. 4.7b.

Combined oscillograms in the absence of a turn-to-turn closure (defectogram) are
shown in Fig. 4.7c.

The presented oscillograms clearly show the difference between the response
signals obtained in the presence and absence of a defect condition. This confirms the
possibility of the suggested diagnostic method to identify turn-to-turn short circuits
of two adjacent turns (a1–a2).

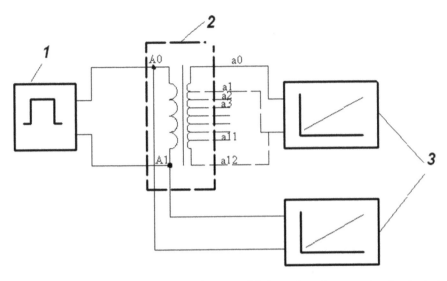

Fig. 4.5 Diagram of a single-phase transformer model: 1—probe pulse generator; 2—physical model of a power transformer based on a PMA-6100 magnetic starter; A0, A1—leads from the PMA-6100 coil, a0, a1–a12—leads from an additional coil; 3—TDS 1012B oscilloscopes

Fig. 4.6 Oscillogram of a probe pulse

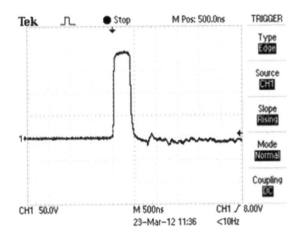

4.2.3 Passage of Probe Pulses Through the Windings of the Physical Model of a Transformer

4.2.3.1 Passage of Probe Pulses Through the LV Winding

The appearance of the low voltage (LV) winding is shown in Fig. 4.4, and the electrical circuit of the LV winding is shown in Fig. 4.8.

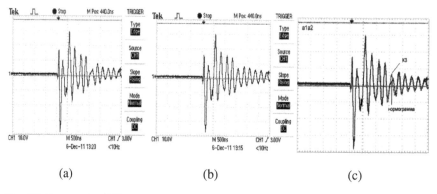

Fig. 4.7 Response oscillograms: **a** response in the absence of a defect, **b** response when two adjacent turns a1 and a2 are closed (see Fig. 4.5), **c** combined response in the absence of a defect and at closure of two adjacent turns a1 and a2

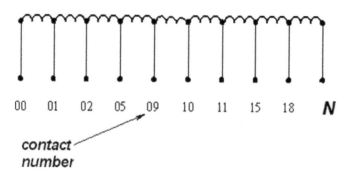

Fig. 4.8 Electrical circuit of the LV coil

A single pulse from the Nanotest-1 generator was applied to the low-voltage winding of the physical model of a transformer. A Tektroniks TDS 2012B oscilloscope was used to measure the probe pulse and response signals. A probe pulse with amplitude of about 200 V and duration of 400 ns was applied to the terminals of the LV winding (20 turns); in this case, the response signal was recorded on the taps corresponding to 5, 10, 15 and 20 turns of the LV winding, counting from the zero potential of the winding N. The corresponding response signals are shown in Fig. 4.9.

These oscillograms were obtained as normograms, so that in the subsequent modeling of various defects in the LV winding, these oscillograms served as a standard. The next series of measurements was performed in the case of a defect simulated in different parts of the LV winding, namely, a short circuit of one or several turns in different parts of the LV winding. In Fig. 4.10, as an example, combined oscillograms from the 10th turn of the LV winding are shown in the absence of a short circuit (curve 1) and with a short circuit of three turns at the beginning of the winding (curve 2).

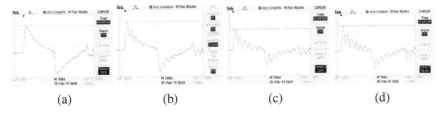

Fig. 4.9 Response signal from the LV winding from: **a** turn 5, **b** turn 10, **c** turn 15 and **d** turn 20

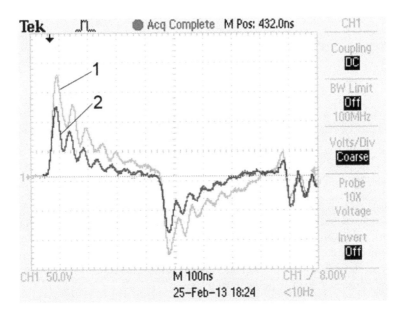

Fig. 4.10 Response signal from the 10th turn of the LV winding: 1–no closed turns in the LV winding, 2—short-circuit of three turns at the beginning of the winding

The above oscillograms show that in the presence of a turn-to-turn SC at the beginning of the winding (the first three turns at the beginning of the winding are closed), the amplitude of the first peak decreases, and the response curve oscillations decay more rapidly than the curve with a winding without short-circuited turns. With an increase in the number of closed turns of the winding, the first peak of the response amplitude decreases even more significantly, while the number of oscillations and their amplitude decrease noticeably. Manifestations of this effect have been confirmed by combined oscillograms on the 15th turn of the LV winding, Fig. 4.11. Here, curve 1 presents a short circuit at the beginning of the winding (the first three turns are closed at the beginning of the winding, and two turns (10–12) are closed in the middle of the winding); curve 2 presents a short circuit at the beginning of the winding plus two turns in the middle of the winding (turns 10 and 12) and three turns at the end of the winding (turns 18, 19 and 20).

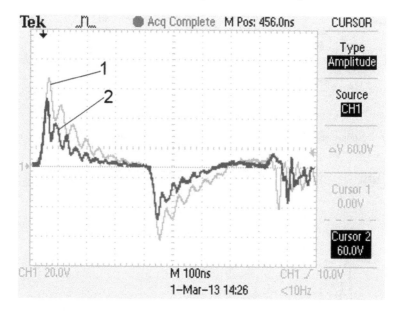

Fig. 4.11 Response signal from the 15th turn of the LV winding: 1—closed turns 1–3, 10–12; 2—closed turns 1–3, 10–12, 18–20

It should be noted that such a technique—applying a probing pulse and taking a response at the same winding—is not typical for traditional diagnostic procedures. An important advantage of the modernized technique is the ability to trace the sequential change in the parameters of response signals (shape, structure) with a sequential increase in the degree of winding defectiveness.

4.2.3.2 Passage of Probing Pulses Through the High-Voltage Winding

In this series of experiments, the HV winding consisted of 120 turns and contained 12 taps, which made it possible to arrange various combinations of turn-to-turn short circuits. The LV winding was located inside the HV winding and was located on the **B** phase bar of the magnetic conductor. The probing pulse acted on the HV winding between the extreme upper and lower terminals. The lower terminal was grounded and connected to the lower yoke. To register the signals, two oscilloscopes were used: a two-beam Tektronix TDS 1012 V, used to record two probing pulse signals: a pulse at the output of the probe pulse generator and a pulse at a matched load, and a four-beam Tektronix TDS 2024 V used to record response signals. The response signals were recorded on four taps of the LV winding: on turns 5, 10, 15 and on the entire LV winding (on the 20th turn). All response signals from the LV winding were recorded with a four-channel oscilloscope. The first (upper) beam recorded the response from the 5th turn of the winding, and the fourth (lower) beam, from the entire winding (at full length). The second and third beams recorded the response signals for turns 10 and 15 turns of the winding, respectively. The physical model for this series of experiments is shown in Fig. 4.12. Figure 4.13 shows an image of the HV winding with taps and a simulated SC mode in the upper part of the winding. The oscillogram of a probe pulse is shown in Fig. 4.14.

Fig. 4.12 Physical model with windings on the phase *B* bar

Fig. 4.13 HV winding: SC mode simulated at the top of the winding

Fig. 4.14 Oscillogram of a probe pulse

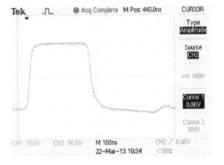

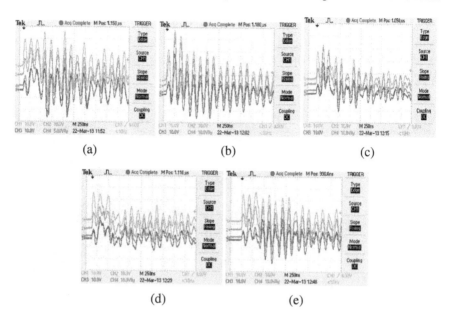

Fig. 4.15 Oscillograms of response signals at different SC points in the HV winding: **a** winding without defects (no closed turns), **b** the first 12 turns are closed, **c** turns 12–24 are closed, **d** turns 60–72 are closed, **e** turns 108–120 are closed

Figure 4.15 shows the oscillograms of response signals when the turns are closed in different locations of the winding.

From the given oscillograms, it follows that the presence of a defect changes the parameters of response signals: the oscillation period, the response signal components, the amplitude of peaks and their repetition rate. This change can be observed without any special software for comparison. The sequential movement of a defect along the winding is reflected as a characteristic change in the shape and structure of the response signals. A decrease in the amplitude of peaks and their characteristic "compression" are signs of a defective winding condition.

4.2.4 Detection of Radial and Axial Displacements of Winding Turns

Radial displacement of winding turns is one of the types of mechanical defects. To simulate a diagnostic procedure for identifying this type of defect, the following experiment was performed. The radial displacement was simulated on the HV winding by introducing a dielectric insert between the winding body and the turns. The turn displacement was 3 mm. Coils 12–15 were displaced. The external view of the HV winding with a simulated defect is shown in Fig. 4.16. The probe pulse

Fig. 4.16 Simulated defect "radial displacement of winding turns"

generator was connected to the input of the HV winding. The response signal was recorded at the LV winding. The oscillograms of response signals in the absence and presence of a defect are shown in Fig. 4.17.

The first part of the response signal is practically identical with the normogram. The difference in the waveforms appears in the second part of the curve. A change in the waveform at a certain section indicates a change in the winding. Similar deviations of the response signal from the normogram are observed when the turns of the HV winding are displaced in the axial direction. Figure 4.18 shows the combined oscillograms of a defect-free winding (curve 1) and a winding with a shift of several turns in the lower part (curve 2).

The oscillograms shown in Figs. 4.17 and 4.18 demonstrate that the deviations of a normogram from response signals at a radial displacement of turns or at their axial displacement are but slightly different, so that for an objective assessment of the difference between oscillograms a special program for digital data processing has been developed (see in Appendix 1).

Fig. 4.17 Oscillograms of response signals: 1—no defect, 2—buckling of turns according to Fig. 4.16

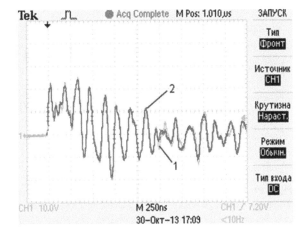

Fig. 4.18 Oscillograms of
response signals: 1—no
defect, 2—displacement of
several turns in the axial
direction in the lower part of
the HV winding

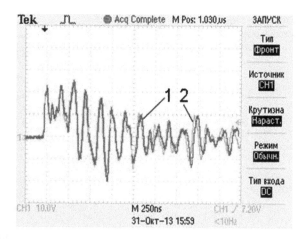

4.2.5 *Regularities of Response Formation for Various Methods of Connecting the Windings*

Two winding connection schemes have been experimentally investigated: "star-star" and "star-delta"

A.   *"Star-star" connection of windings*

A rectangular pulse from the Nanotest-1 generator was applied to the high-voltage input of one of the phases. The response pulses (hereinafter called the response) in the low-voltage winding were recorded using an oscilloscope. It is experimentally shown that with such a connection of the windings, a turn-to-turn circuit in the low-voltage winding is determined by the response in the corresponding low-voltage winding. As an example, Fig. 4.19 shows an oscillogram of the response at the LV winding of phase *b* in the absence of a defect (Fig. 4.19a) and also one with a short circuit of

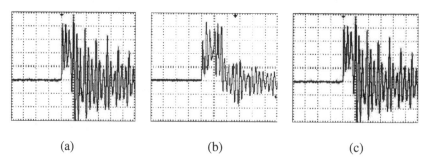

       (a)   (b)   (c)

Fig. 4.19 Response from the LV winding ("star-star" connection): **a** response from phase *b* in the absence of defects, **b** response from phase *b* when several turns of the LV winding of phase *b* are closed, **c** response from phase *a* when closing several turns of the LV winding of phase *b* (the scale of all the oscillograms is the same on both axes)

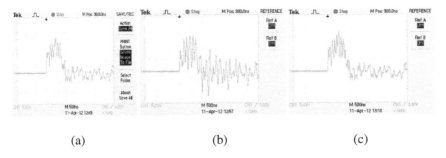

(a)   (b)   (c)

Fig. 4.20 Response from the LV winding ("star-delta" winding connection): **a** response from phase *b* in the absence of defects, **b** response from phase *b* when several turns of the LV winding of phase b are closed, **c** response from phase *b* at closing of several turns of the LV winding of phase *a*

several turns of the low-voltage winding of phase *b* (Fig. 4.19b). It can be seen that the response in the LV winding of phase *b* has changed. In the presence of several short-circuited turns in the LV winding of phase *b*, when a probing signal is applied, for example, to the HV winding of phase *A* and is recorded in the LV winding of phase *a*, the response is almost unchanged in comparison with the response of a defect-free winding (Fig. 4.19c).

In Fig. 4.19 the responses show that when the windings are connected according to the "star-star" type the method is sensitive only in the case of comparing the responses on phases of the same name.

B.   *Connection of windings according to the "star-delta" scheme*

The experimental technique is similar to that used in "star-to-star" connections. The probing pulse was applied to one of the HV windings, and the response was recorded on one of the HV windings. Figure 4.20 shows responses at the HB windings when a probe pulse is applied to the HV winding of phase *B*.

The results show that when the windings are connected according to the "star-delta" scheme and a probe pulse is applied to one of the windings connected as a "star", response is observed on all the windings connected as a "delta". This makes it difficult to identify the damaged winding, because one does not know in advance what response should be in a damaged and undamaged winding. In such cases, it is advisable to carry out mathematical simulation in order to understand how a short circuit in one of the windings should affect the form of the response in the other windings connected as a "delta".

4.2.6   Possibilities of Diagnostic Procedures Under Operating Voltage: Experimental Research

In order to investigate the possibility of monitoring the state of transformer windings under operating voltage, a separating K15-4 capacitor was installed between the probe pulse generator and one of the transformer phases. The parameters of the capacitor are: rated voltage of 40 kV, capacity of 2200 pF. The same capacitor was used in the measuring circuit. Thus, the probe pulse generator and measuring equipment were isolated from the transformer windings, which can be implemented in practice to monitor the state of transformer windings in the operating mode under the operating voltage of a transformer. The connection diagram of the diagnostic equipment with the transformer is shown in Fig. 4.21. Figure 4.22 shows two response signals obtained under the same experimental conditions in the presence and absence of coupling capacitors C and $C1$. The response signals in both cases are completely identical, which shows that there is no distortion of the diagnostic procedure due to the coupling capacitors. Figure 4.23 shows an oscillogram of the response signal in the case of a short circuit in the LV winding of phase B. The experimental results unambiguously demonstrate the fundamental possibility of implementing a diagnostic procedure under operating voltage.

The capability to carry out a diagnostic procedure under operating voltage is an important result of this research, since monitoring the condition of a transformer

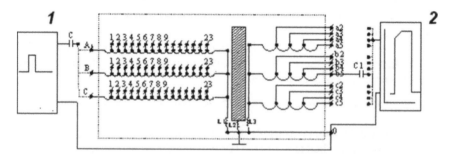

Fig. 4.21 Circuit for monitoring the state of windings under operating voltage: 1—probe pulse generator, 2—registration system, C and C1—coupling capacitors

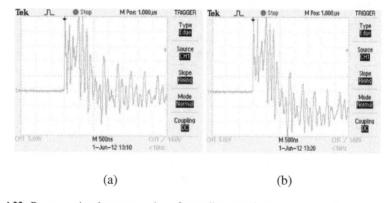

(a)   (b)

Fig. 4.22 Response signal: **a** no capacitors; **b** coupling capacitors are connected

Fig. 4.23 Response signal in the presence of an SC in the LV winding of phase *B*

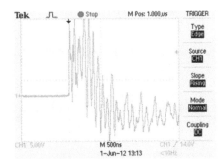

without disconnecting and unbussing, i.e., in the process of operation, simplifies this procedure and reduces the costs involved.

4.2.7 Influence of Probe Pulse Parameters on Diagnostic Sensitivity Investigated

For a more reliable assessment of the capacities of the proposed technique, additional experimental research was performed, consisting in the fact that in the region of the low-voltage coil, a conductor about 5 cm long was soldered to one of the turns located in the middle of the winding, in parallel to the latter, as shown in Fig. 4.24.

The presence of this conductor introduced an additional insignificant electrical capacitance between two adjacent turns of the LV winding and relative to the surrounding space. A probing pulse of rectangular shape with duration of 400 ns was applied to the input of the LV winding (see Fig. 4.14). A response signal from this LV winding was recorded in the absence and presence of the specified conductor

Fig. 4.24 Additional capacitance near the LV winding

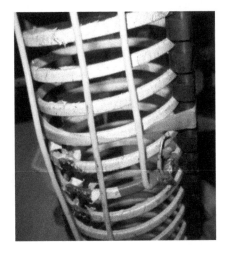

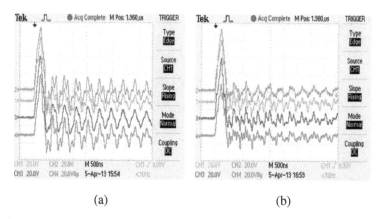

(a)   (b)

Fig. 4.25 Response obtained in the experiment with an additional capacity: **a** without an additional capacity, **b** with an additional capacity

(additional capacitance). The response without a conductor is shown in Fig. 4.25a, and with a conductor, in Fig. 4.25b.

The presented oscillograms show a significant difference in the shape of the response signals. The oscillation period, shape, amplitude and general appearance of the curves have changed substantially. Such a difference in the shape of signals after a 5 cm long wire has been introduced into the winding region confirms the high sensitivity of the proposed diagnostic method based on the use of short probing pulses. To reveal the influence of the shape of a probe pulse on the sensitivity and the general scheme of the diagnostic procedure, an experiment was carried out in which two types of pulses were used: a "bell" shaped pulse with different values of rise time and a rectangular pulse. The shape of the probe pulse was changed by installing capacitors with capacities of 2200 and 1600 pF at the output of the probe pulse generator. The pulse was applied to the input of the HV winding. The response signals were recorded at the LV winding on taps corresponding to turns 5, 10, 15 and 20 of the LV winding. Figure 4.26, shows the form of a probe pulse, and Fig. 4.26b shows the corresponding response.

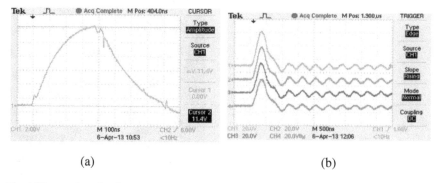

(a)   (b)

Fig. 4.26 **a** probe pulse, **b** response

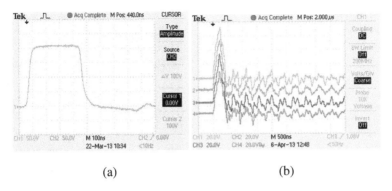

(a)   (b)

Fig. 4.27 **a** probe pulse, **b** response

Figure 4.27 shows the form of a rectangular probe pulse, and Fig. 4.27b shows the corresponding response.

One can see that the response to a rectangular pulse contains a wider spectrum of harmonics than the response to a pulse with a shallow front; this has affected the sensitivity of the diagnostic procedure, which is significantly higher in the case of rectangular pulses. According to the results of theoretical research and the practice of using the pulsed diagnostic method, the duration of a probe pulse is an important parameter. Experiments have been performed to determine the effect of pulse duration on the sensitivity of the diagnostic procedure. The duration of a probe pulse was adjusted by changing the length of the generator cable. The probing pulse was applied to the input of the HV winding. The response signals were recorded on the taps of the LV winding corresponding to turns 5, 10, 15 and 20. Two conditions were simulated: (a) a defect-free state of the winding, (b) a short circuit in the region of the lower part of the HV winding. Figure 4.28 shows a fragment of the winding with closed turns 96 and 108.

Response signals were measured at probe pulse duration of 350, 77, 41, and 10 ns. As an example, Fig. 4.29 shows oscillograms for probing pulses with duration of 350 and 10 ns.

One can see that an increase in the probe pulse length makes the response less saturated with harmonics. This means that the sensitivity of the diagnostic procedure decreases with an increasing pulse length.

To study the influence of the rise time of a probing pulse on the sensitivity of the diagnostic procedure, a rectangular pulse was applied to the high-voltage winding of the transformer (Fig. 4.30, oscillogram 1), the response pulse was taken from the low-voltage winding of the transformer (Fig. 4.30, oscillograms 2 and 3). The windings were unloaded (bus-bar transformer model). Pulses with a rise front time of 400 and 25 ns were used. The SC mode was simulated by electrical connection between turns 96 and 108 in the winding (see Fig. 4.28). Oscillograms of response from the low-voltage winding of the transformer in the absence of winding defects (Fig. 4.30, oscillograms 2) and in the presence of a short circuit at the end of the HV winding

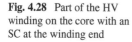

Fig. 4.28 Part of the HV
winding on the core with an
SC at the winding end

(Fig. 4.30, oscillograms 3) are different from each other the more significantly, the
shorter the front of a probe pulse.

The obtained results can be explained by the fact that pulses with a shorter
leading edge excite circuits with a higher natural frequency. Thus, using pulses with
a steep edge for probing enables one to track even minor changes in the geometry of
transformer windings, which is problematic when using pulses longer than 500 ns.

4.2.8 Research of a Power Transformer Using a Mathematical Model

An algorithmic implementation of our mathematical model is presented in the
MATLAB Simulink software integrated environment. The model consists of two
main units and a number of auxiliary ones. The main units are the power supply
unit and the transformer winding simulation unit. The power supply unit is designed
to form a probing pulse in the form of an algebraic sum of Fourier components
with corresponding amplitudes and phases. The transformer winding simulation unit
shown in Fig. 4.31 represents the transformer windings in the form of a chain diagram
of links connected in series.

A circuit equivalent of a winding turn used in the mathematical model is shown
in Fig. 4.32.

Figure 4.33 shows an example of a simulated oscillogram for a probe pulse and
a corresponding spectral composition.

Figure 4.34 shows an example of response on the LV transformer winding when
probing with the pulse shown in Fig. 4.33.

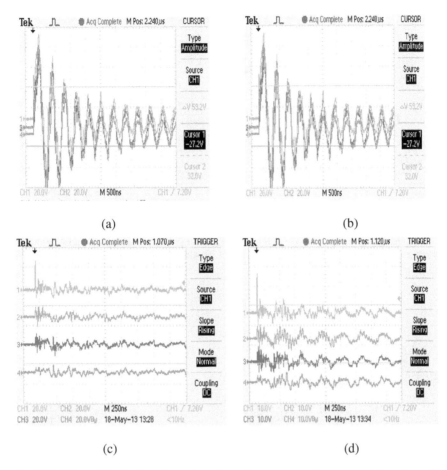

Fig. 4.29 Effect of the duration of a probe pulse on the response: **a, b** response at the probe pulse duration of 350 ns; **c, d** response at the probe pulse duration of 10 ns; **a, c** defect-free winding; **b, d** closure of several turns at the end of the HV winding

The developed mathematical model of a power transformer makes it possible to simulate winding defects (turn-to-turn circuits) in different locations with a different number of closed turns and their various combinations. The response signals obtained by applying a simulated probe pulse are in satisfactory agreement with the response signals obtained in experiments using the physical model. This result makes it possible to consider the developed mathematical model as adequate and use it for improvement of the diagnostic procedure [6, 7].

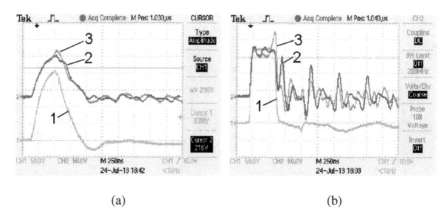

(a)   (b)

Fig. 4.30 Comparison of response pulses at different rise times of a probing pulse: **a** 400 ns, **b** 25 ns: 1 is the shape of a probe pulse on the HV winding; 2 is the shape of a response pulse on the LV winding without defects; 3 is the shape of a response pulse on the LV winding when closing turn 96 with turn 108

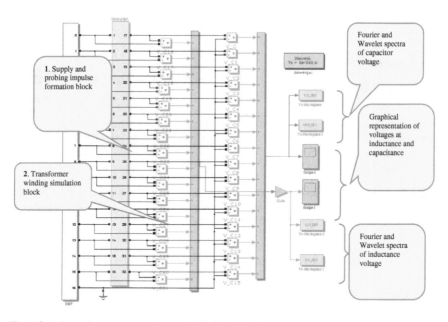

Fig. 4.31 General appearance of the model in block design

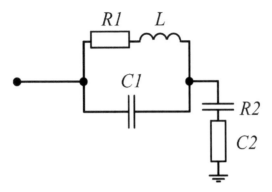

Fig. 4.32 Electrical circuit equivalent of a winding turn

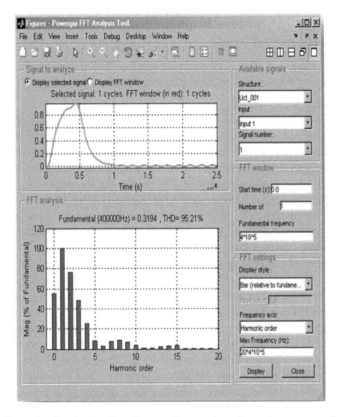

Fig. 4.33 Example of a simulated oscillogram for a probe pulse and a corresponding spectral composition

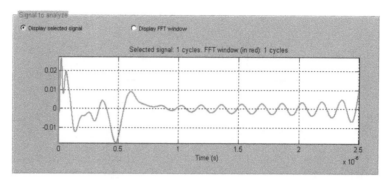

Fig. 4.34 Response on the LV transformer winding when probing with the pulse shown in Fig. 4.33

4.3 Diagnostic Complex Implementing the PD Technology in Nanosecond Mode

4.3.1 Object Description and Experimental Procedure

To determine the level of sensitivity, reliability and efficiency in the use of a diagnostic complex under industrial conditions, research was conducted on serviceable and defective transformers of the same brand manufactured at the same enterprise and operated in practically the same conditions by using two methods: the one developed by the authors of this monograph and a method being the most similar to ours in analyzing the frequency characteristics. Three-phase oil-filled power transformers with a capacity of 160 kVA, 10 kV were located at a specialized repair shop of an energy enterprise in one of the northern energy systems of the Russian Federation. The main data of the examined transformers are shown in Table 4.1.

The scheme of monitoring the condition of transformer windings, including a generator of probing pulses of the nanosecond range, oscilloscopes to control the parameters of probe pulses and response signals, as well as the location of the phases of the LV and HV windings, is shown in Fig. 4.35.

The number of turns on the HV winding is 1312; the number of turns on the HN winding is 160. The control procedure consisted in applying to one of the windings a sequence of probe pulses with a duration of 520–25 ns and an amplitude of 200 V, as well as in recording response signals corresponding to a transient process from the other winding. The measuring procedure was completely identical to that used in experiments on our physical model of a transformer and is described in detail in Chaps. 2, 3 and 6. In the case of using the FRA method, the FRAX 150 instrument system was used whose screen displayed amplitude-frequency characteristics pre-processed using software and hardware.

Below we present the results of comparing the two response signals for different winding conditions. The comparison of measurement results was carried out using a special program developed for the specified purpose.

Table 4.1 Data of object under survey

Option no.	Name of option	Data of object under survey
1	Transformer location	Ishim repair plant "Tumenenegro", Ishim, Tumen Region, Russia
2	Name of object	Transformer
3	Type of transformer	Three-phase, total capacity 160 kVA, 10 kV, oil-filled, distributed grid transformer
4	Date of issue, Production plant	"Altay transformer", Barnaul, Russia
5	Start date of operation	1993
6	Winding connection scheme	Y/Y_0
8	Transformer capacity	160 kVA
9	Operating voltage of HV winding	10 kV
11	Operating voltage of LV winding	0.4 kV
13	Cooling type	Oil type
15	Problem condition during operation	HV winding damaged by lightning. Object repaired at specialized plant.
18	Total mass	800 kg
19	Oil mass	225 kg

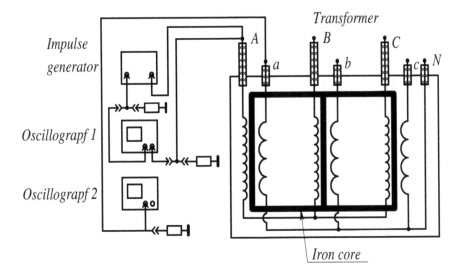

Fig. 4.35 Scheme of experiment at condition control using the pulsed method

4.3.2   Choice of Efficiency Criterion

One of the main efficiency indicators of a condition monitoring method is sensitivity. The sensitivity, in turn, can be estimated using the concept of the average integral of voltage pulse. The latter was determined by averaging the integral of voltage pulse for 27 oscillograms of a defect-free transformer. The average integral of voltage pulse on the HV winding was $8.75 \times 10\,5$ (Vb). The deviation from the average of an integral of voltage pulse did not exceed 2% from pulse to pulse. This may be considered as an error (noise) in determining an integral of voltage pulse on the HV winding. The average integral of voltage pulse on the LV winding when processing 27 oscillograms was $7.52 \times 10\,5$ (Vb), and the deviation from the average was 1.6%. The deviation of an integral of voltage pulse from the average for a response pulse over three oscillograms ranged within 3-4%. Thus, it can be assumed that during a probing the error (noise) lies within 3-4% of an integral of voltage pulse. If the integrals of voltage pulse for the normogram and defectogram differ by more than 5% (excess over the noise), then this deviation can be considered as an indication of the presence of a defect condition in the transformer winding.

4.4   Effectiveness Comparison for the FRA and PD Technologies

To compare the sensitivity of the proposed method and the FRA method, a defect of the "turn-to-turn short circuit" type was arranged in the HV winding of phase C. The appearance of a defective winding is shown in Fig. 4.36.

Fig. 4.36 Defect of the "turn-to-turn short circuit" type on the HV winding of phase C

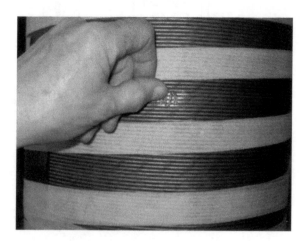

Before pulsed probing, the winding condition was tested by resistance measurement. The resistance before short-circuiting three turns was, respectively, **A-C**—0,1046 O, **B-C**—0,1042 O, and after short-circuiting, **A-C**—0,1046 O, **B-C**—0,1043 O (measured with a PTF-1 milliohmmeter). Thus, the resistance measurement did not detect the presence of three short-circuited turns.

4.4.1 Diagnostics by Nanosecond Pulses

Figure 4.37 shows, as an example, the result of comparing the response signals to a nanosecond pulse: a blue curve in the absence of defects, a red curve in the case of a turn-to-turn short circuit. In all the cases, the neutral terminal is grounded, and the rest of the winding leads are in idle mode.

Processing oscillograms according to the developed program shows that the integral of the difference between the voltage pulses of the normogram and defectogram is 3.4×10^{-6} in the time interval from 0 to 2.2×10^{-6} s, the integral of voltage pulse of the normogram in the same interval equals 1.01×10^{-5}. The integral of the difference between the voltage pulses of the normogram and defectogram, as a percentage of the integral of transformer response voltage pulse, equals 34%. This result, according to the logic given in Sect. 4.1, shows that the use of the proposed oscillogram processing technique enables one to register a significant difference between the normogram and defectogram when 3 turns of the HV winding are short-circuited.

Table 4.2 shows the results of processing oscillograms according to the above method for various combinations of the probe pulse and response action.

Resistance measurement of the HV windings using an ohmmeter did not reveal any difference between a defect-free winding and a winding with three short-circuited turns in a total number of turns on the HV winding being 1312. Using the method of nanosecond pulses, a difference is recorded in the range from 35% to 55%, which

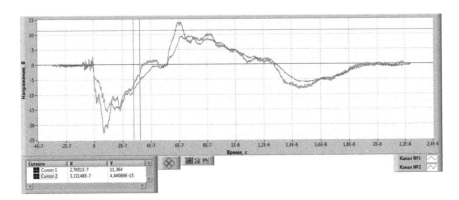

Fig. 4.37 Oscillograms of probe pulses applied to the winding of phase *A* and the response signal recorded in phase *a*

Table 4.2 Difference between normograms and defectograms for various combinations of applying the probe pulse and response

No.	Winding acted upon by probe pulses	Winding of response pulse recording	Integral of difference between response voltage pulses of a defect-free winding and a winding with 3 short-circuited turns of phase C, Wb	Integral of voltage pulse difference, in % of the integral of response pulse for a defect-free winding
1	A	B	8.7×10^{-7}	45
2	A	C	8.9×10^{-7}	47
3	B	C	9.3×10^{-7}	47
4	C	A	8.7×10^{-7}	44
5	C	C	3.5×10^{-6}	55

depends on a combination of the winding acted upon by the probing pulse and the winding on which the response pulse is recorded. One can see that the highest sensitivity is achieved in a combination when the probe pulse applied to the HV winding and the response pulse received from the LV winding are recorded from a pair of windings of the same phase. For example, if the defect is located in the winding of phase C, then the largest integral of the difference of response voltage pulses is obtained when a probe pulse is applied to the HV winding of phase C, and the response is recorded from the LV winding of phase C.

4.4.2 Diagnostics Using FRAX-150

Figures 4.38 and 4.39 show the results of the above measurements, albeit carried out using the FRA method.

The results show that the largest sensitivity is achieved in the combination when a probe voltage and a response are recorded on a pair of windings of the same phase, i.e., the same as in the case of using nanosecond pulses. Differences in the amplitude-frequency response are observed in the frequency range from 200 Hz to 300 kHz. The FRA method and the nanosecond pulse method are sensitive to short circuits of several turns in the HV winding and allow them to be detected.

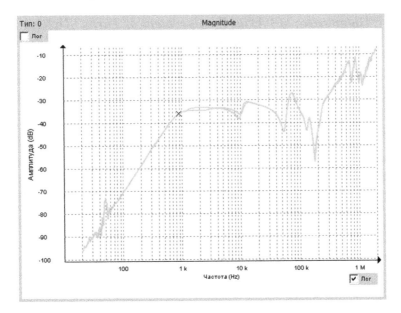

Fig. 4.38 A probe pulse is applied to the winding of phase A, the response signal is recorded on phase a

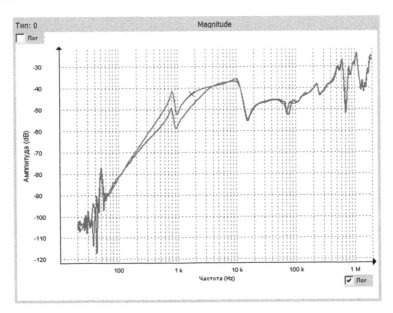

Fig. 4.39 Probe pulse is applied to the winding of phase *A*; the response signal is recorded in phase C

4.4.3   Diagnostics of "Axial Displacement of Turns" Defetcs in the HV Winding of Phase a by Probing with Nanosecond Pulses

To assess the diagnostic sensitivity in the presence of the "axial displacement of turns" type of defects, a defect was created on the winding of phase A covering 14 turns (described in Sect. 4.2.7). The appearance of the HV winding affected by such a defect is shown in Figs. 4.40a, b.

Below we present the results of comparing the signals of response to the action of a nanosecond pulse in the absence of defects (the blue curve) and the signals in the case of axial displacement of turns (the red curve), with various combinations of the probe pulse location and the response registration location. In all cases, the neutral terminal is grounded; the rest of the winding leads are in idle mode. Figure 4.41 shows combined response oscillograms obtained from the windings of phase A.

Table 4.3 shows the results of processing oscillograms similar to those shown in Table 4.2, albeit in the case of "axial displacement of turns" in the HV winding of phase A. Signal processing was carried out according to the above method: the difference of the integrals of response pulses for a defect-free winding and a winding with displaced turns (Fig. 4.41) was compared with the integral of a response pulse for a probe pulse.

One can see a significant difference between the response pulses of normograms and defectograms. It is difficult to set a sensitivity limit in this case, because the situation involving a displacement of turns leads to a change in the capacitance ratios in almost all of the phases and depends significantly on the geometric dimensions of

(a)   (b)

Fig. 4.40 HV winding of phase A with an "axial displacement of turns" defect: frontal **a** and profile **b** views of the coil

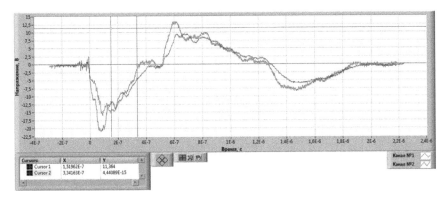

Fig. 4.41 A probe pulse is applied to the winding of phase *A*, the response signal is recorded on phase *a*

Table 4.3 Difference between normograms and defectograms for various combinations of the impact of a probing pulse and a response for an axial displacement of 14 turns in the HV winding of phase *A*

No.	Winding acted upon by probe pulses	Winding of response pulse recording	Integral of difference between response voltage pulses of a defect-free winding and a winding with 14 axially displaced turns of phase *A*, Wb	Integral of voltage pulse difference, in % of the integral of response pulse for a defect-free winding
1	A	A	3.4×10^{-6}	38
2	A	C	9.3×10^{-7}	48
4	C	A	9.5×10^{-7}	48

transformer coils and the number of turns shifted in the axial direction. Apparently, for each type of transformer, this should be considered as an individual parameter.

4.4.4 Diagnostics of "Axial Displacement of Turns" Defects in the HV Winding of Phase a by FRA Method Using FRAX-150 Device

Figure 4.42 shows the results of diagnostics by the FRA method in the following case: the probing signal is applied to the HV winding of phase *A*; the response signal is recorded on the LV winding of phase *a*. The signals are superimposed for a defect-free winding and for a winding with axially displaced 14 turns in the HV winding of phase *A*.

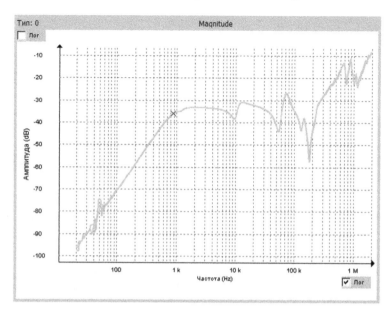

Fig. 4.42 Probe pulse is applied to the HV winding of phase *A*; the response signal is recorded on phase *a* (the signals are superimposed for visual comparison of their shape)

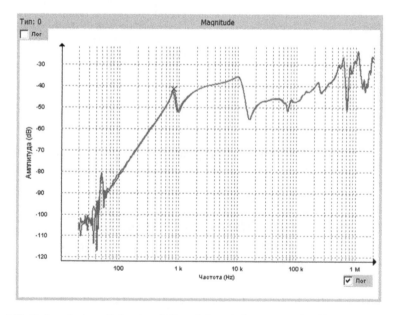

Fig. 4.43 Probe pulse is applied to the winding of phase *A*; the response signal is recorded on phase *c*

Figure 4.43 shows the results of measurements by the FRA method in the following case: the probing signal acts on the HV winding of phase **A**, the response signal is taken from the LV winding of phase **c**. The signals are superimposed for a defect-free winding and for a winding with axially displaced 14 turns of the HV winding of phase **A**. Figures 4.44, 4.45 and 4.46 show similar results of measurements by the FRA method in other cases of exposure to probing pulses and corresponding

Fig. 4.44 Probe pulse is applied to the winding of phase **B**; the response signal is recorded on phase **b**

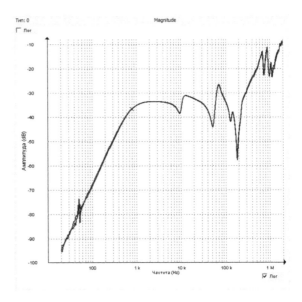

Fig. 4.45 Probe pulse is applied to the winding of phase **B**; the response signal is recorded on phase **c**

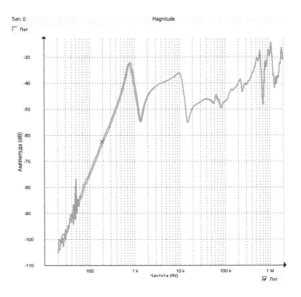

Fig. 4.46 Probe pulse is
applied to the winding of
phase C; the response signal
is recorded on phase c

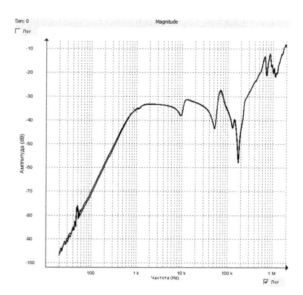

response signals. In all the cases, the pulses affected the defect-free winding and the winding with 14 turns of the HV winding of phase A shifted in the axial direction.

One can see that the difference between the amplitude-frequency characteristics is difficult to grasp. Similar patterns are observed with other combinations of the schemes. It is extremely difficult to draw a conclusion about the presence of a defective winding condition on a basis of the above pictures, and therefore the probability of an erroneous diagnosis is unacceptably high Based on the results of studies, it should be recognized that probing with nanosecond pulses and using the FRA method react with approximately the same sensitivity to turn-to-turn circuits, but the former method is more sensitive to the displacement of turns in the high-voltage winding of the transformer. Thus, the method of nanosecond pulses allows detecting a displacement of 14 turns in the high-voltage winding, whereas such a defect cannot be detected by the FRA method [8, 9].

4.5   Technology of Single-Stage Pulsed Defectography

This section presents the results of original research, the purpose of which is to further improve the efficiency of monitoring the condition of windings. The first stage is the use of a one-stage technology based on the use and subsequent analysis of only a probing pulse of nanosecond duration applied to a winding under study. In this case, the recording and processing of response signals from adjacent windings is not carried out. The second stage is the use of switching pulses of a high-voltage power grid as probing ones.

4.5.1 Winding Condition Monitoring While Using a Single Stage

The present approach to the implementation of pulsed defectography involves the use of only the first stage of diagnostic procedures using the LVP method: the supply of a probe pulse to a winding under study. This should significantly reduce the measurement time, as well as the negative impact of additional electrical circuits and measuring instruments, reducing, as a result, the likelihood of an erroneous diagnosis.

The low-voltage pulsed probing technology used in current practice, based on a change in the amplitude and frequency of the signal under investigation, makes it possible to make only a qualitative assessment of the measurement results and use it in an attempt to judge about the presence or absence of damage in transformer windings. This technique is complicated and not quite reliable. Spectral analysis of signals allows one to eliminate or at least minimize these disadvantages, i.e., to analyze much more accurately and thoroughly the processes in transformer windings. Using the amplitude-frequency characteristic (AFC), it is possible to estimate a change in the frequency spectrum due to various types of damage in a transformer. To construct the frequency response, a numerical spectral analysis based on a discrete Fourier transform (DFT) was used. The Fourier series of pulses was decomposed using the *MathCAD* software package according to the formula

$$F_n = \frac{1}{N} \sum_{k=0}^{N-1} S_k \exp\left(-j\frac{2\pi nk}{N}\right), \tag{4.1}$$

where: N is the number of signal values measured over a certain period;

S_k are signal values measured at discrete points in time;

F_n are complex amplitudes of sinusoidal signals that make up the original signal (denoting both the amplitude and the phase);

k is a time index of input counts, $k = 0,...,N-1$;

n is an index used in the frequency domain ($n = 0,...,N-1$).

In the course of spectral analysis, the original pulse was decomposed into 50 harmonics, after which, using the inverse discrete Fourier transform, this pulse was reconstructed using a different number of harmonics. The results of an inverse DFT are shown in Figs. 4.47 and 4.48.

The graphs of the original and reconstructed impulses show (Fig. 4.48) that the minimum number of components of a Fourier series for which it is necessary to analyze the initial impulse is 50. Otherwise (Fig. 4.47) there is a noticeable difference between the original and reconstructed pulses.

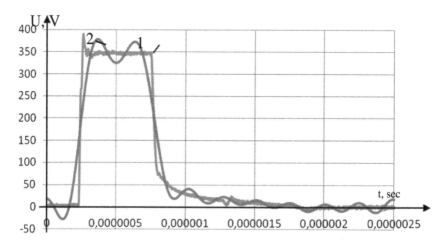

Fig. 4.47 Graph of the original signal (1) and one reconstructed by 10 harmonics using an inverse DFT (2)

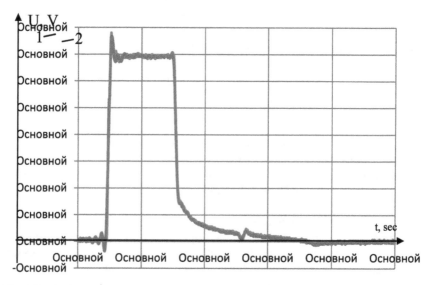

Fig. 4.48 Graph of the original signal (1) and one reconstructed by 50 harmonics using an inverse DFT (2)

4.5.2 Implementation of Single-Stage Defectography

The principle of "classical" defectography is based on the fact that a short rectangular probe pulse of relatively low voltage is fed from a special generator to a winding input, and responses are recorded from the inputs of the other windings.

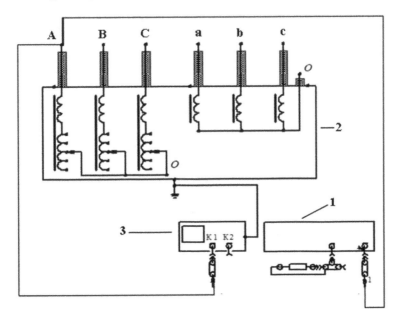

Fig. 4.49 Scheme of single-stage technology implementation in experiments: 1—generator of probing pulses, 2—high voltage 10/0.4 kV transformer, 3—Tektronix TDS 1012 oscilloscope

The principle of single-stage technology involves the supply of only a probe pulse and its registration as a response pulse, with a subsequent spectral analysis [10]. Figure 4.49 shows a diagram for monitoring the condition of windings.

4.5.3 Comparison of Spectra for Pulses Applied to Defect-Free Windings

In order to quantify the difference between signals, it is necessary to determine the number of signal harmonics being investigated. As mentioned above, the minimum number of components of a Fourier series into which it is necessary to decompose the initial pulse is 50. In this regard, the frequency response was decomposed at a frequency from 0 to 50 MHz. The object of study was a three-phase oil-filled power transformer with a capacity of 160 kVA and a rated voltage of 10 kV. The first unit of spectral comparison of probing pulse signals applied to a healthy HV winding of phase A was performed. Figures 4.50 and 4.51 depict the AFC of the signals for which it was found by experiment that the average level for the spectral ratio *SR* has an error within 1–1.2. This may be due to different measurement losses.

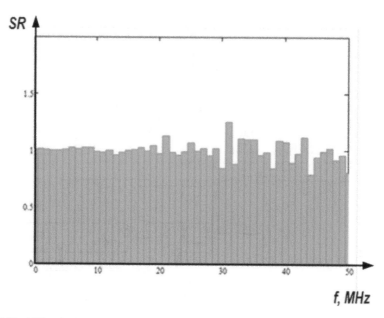

Fig. 4.50 AFC ratio No. 1 for normograms of a healthy HV winding of phase A

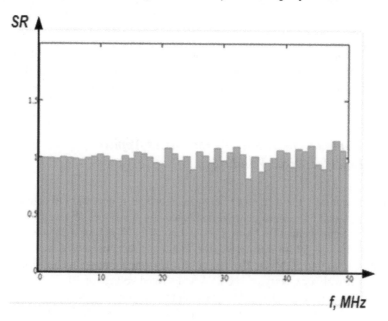

Fig. 4.51 AFC ratio No. 2 for normograms of a healthy HV winding of phase *A*

Consequently, if the AFC of a defectograms has a maximum amplitude value exceeding the margin of error, then such a deviation can be regarded as a sign of damage present in the windings of a transformer under study.

4.6 ON-LINE Monitoring of Winding Conditions

The idea of monitoring the condition of windings in the ON-LINE mode, without using special equipment for generating probing pulses seems to be the most attractive one. This idea can be realized in the conditions of an actual electric power system by using the switching pulses of a high-voltage power grid as probing ones.

Research has been carried out to determine a possibility of monitoring the condition of windings and identifying defects when a transformer is connected to an electric power grid. In this case, a switching pulse arriving at the winding as a result of regular switching plays the role of a probing one. The tested three-phase transformer has two phases in good order, i.e., they are fault-free, whereas the third phase has an artificially created defect. Defect condition was created in the LV and HV windings of phase A of a three-phase oil-filled transformer with a voltage of 10 kV and a capacity of 66 kVA, as shown in Fig. 4.52.

The probe pulse and the response of the windings were recorded using a Tektronix TDS 2012 digital two-channel oscilloscope. *Switching operation "turn on" from the ~ 220 V mains on the LV side*—a response was recorded on the HV side of the same phase using a capacitive voltage divider. The control of the windings is carried out according to the "LV-HV" scheme. Oscillograms of diagnostic results are shown below. The response of the intact pairs, в-B and c-C, differ significantly from the situation for the pair a-A. The response signals were recorded on a "5 μs/division" sweep to observe the initial stage of the transient process (Fig. 4.53), and on a "25 μs/division" sweep, to observe the entire transient process in the winding (Fig. 4.54).

A visual analysis of the obtained oscillograms, without the use of specialized software, makes it possible to ascertain the fact that the responses on the damaged

Fig. 4.52 HV winding damage: axial shift

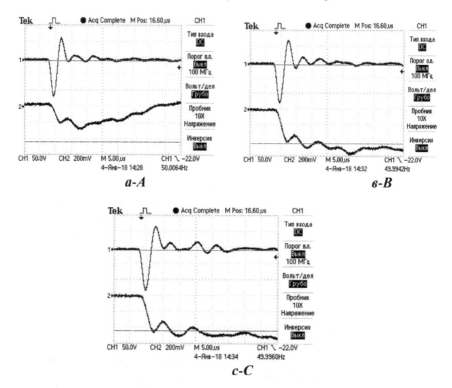

Fig. 4.53 Waveforms of commutation pulse (beam 1) and response (beam 2) for different phases. Oscilloscope sweep is "5 μs/division"

phase differ when the responses on the two undamaged phases are identical and the switching pulse is equal when switched on.

Switching operation "turn on" from the ~ 220 V mains on the HV side. The load was simulated by a rheostat with a maximum resistance of 40 Os. Observation of a response oscillogram according to the "HV-LN" scheme (switching operation "turn on" from the ~220 V mains on the HV side). Oscillograms for various values of resistive loads of the secondary transformer winding are shown in Fig. 4.55. Figure 4.56 shows similar oscillograms of responses at different values of capacitive loads of the secondary transformer winding.

The widespread use of the technology for monitoring the condition of windings under operating voltage will make it possible to bring the diagnostics of transformers to a new technological level. One of the ways to solve the problem of condition monitoring without removing the voltage is to use the switching impulses of electric power grid as probing ones. It has been shown experimentally that such a control makes it possible to detect the axial displacement of winding turns [11, 12].

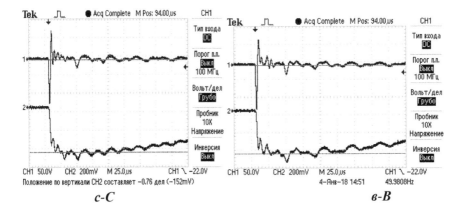

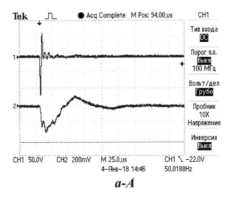

Fig. 4.54 Waveforms of commutation pulse (beam 1) and response (beam 2) for different phases. Oscilloscope sweep is "25 μs/division"

Regardless of the winding control scheme, LV-HN or HN-LV, as well as regardless of the type of load on the LV side, resistive (in experiments, 40 Ohm) or capacitive (in experiments, 0.05 and 0.15 μF), with the same parameters of a switching pulse, the responses from a damaged phase pair (*A-a*) differ significantly from the responses from two serviceable phases (*B-в* and *C-c*). The results of experiments lead to the conclusion that the technology for monitoring the condition of windings based on the use of switching pulses as probing ones is informative and potentially surpasses all the existing methods and tools for diagnosing transformer windings.

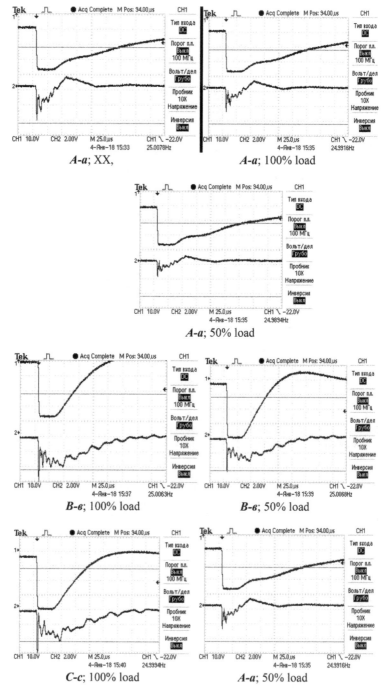

Fig. 4.55 Waveforms of commutation pulse (beam 1) and response (beam 2) for different phases and different resistive loads of the LV winding. Oscilloscope sweep is "25 µs/division"

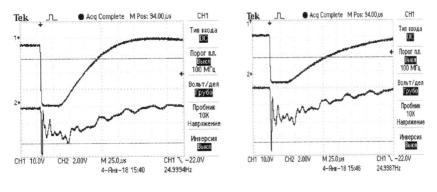

C-a; load – capacitance 0.05 μF;   *C-a*; load – capacitance 0.15 μF

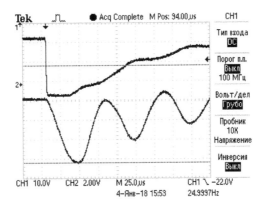

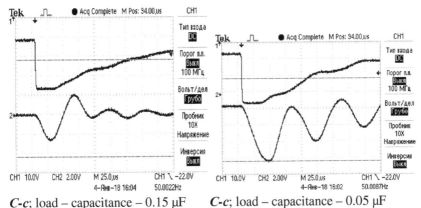

C-c; load – capacitance – 0.15 μF   *C-c*; load – capacitance – 0.05 μF

Fig. 4.56 Waveforms of commutation pulse (beam 1) and response (beam 2) for different phases and different capacitive loads of low voltage winding. Oscilloscope sweep is "25 μs/division"

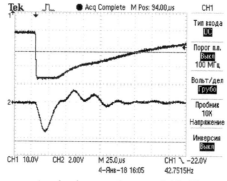

A-a; load – capacitance – 0.15 μF

Fig. 4.56 (continued)

References

1. Lavrinovich VA, Mytnikov AV (2015) Development of pulsed method for diagnostics of transformer windings based on short probe impulse. IEEE Trans Dielectr Electr Insul 22(4):2041–2045. https://ieeexplore.ieee.org/document/7179164
2. Lavrinovich VA, Isaev YN, Mytnikov AV (2013) Advanced control state technology of transformer winding. Int J "Techn Phys Probl Eng" 17(5):94–98
3. Mytnikov A, Lavrinovich A (2019) Further development of transformer winding condition control technology based on pulsed method. Proceedings of 14th international forum on strategic technology, IFOST-2019 October 14–17, 2019 Tomsk, Russia, pp 632–635
4. Lavrinovich VA, Evseeva A, Stepanov I, Mytnikov AV (2017) Development of advanced winding condition control technology of electric motors based on pulsed method. Int J Resour Effic Technol 3(3):232–235. https://reffit.tech/index.php/res-eff/article/view/139
5. Mytnikov A, Lavrinovich A, Strugov V , Saqib M (2021) Development of impulse method for transformer winding condition control technology. IOP conference series: materials science and engineering 1019 (2021) 012024. https://iopscience.iop.org/issue/1757-899X/1019/1/012024
6. Lavrinovich VA, Isaev YN, Mytnikov AV (2014) Modeling of mechanical stress process of transformer winding caused by short circuit currents. Int J "Techn Phys Problems Eng" 19(6):59–63
7. Lavrinovich VA, Isaev YN, Mytnikov AV (2014) Modeling of state control procedure of power transformer winding by short probe pulses. Int J "Techn Phys Probl Eng" 18(6):145–147
8. Lavrinovich VA, Hongda L, Mytnikov AV (2016) Advanced technology of transformer winding condition control based on nanosecond probing impulse. Int J Resour Effic Technol 2(3):111–117
9. Lavrinovich AV, Mytnikov AV (2020) Diagnosticheskiy kompleks dlya effektivnogo kontrolya sostoyaniya obmotok vysokovol'tnykh transformatorov [Diagnostic complex for effective monitoring of the condition of high-voltage transformer windings]. Izvestiya Tomskogo politekhnicheskogo universiteta. Inzhiniring georesursov – Bulletin of the Tomsk Polytechnic University. Engineering of georesources 331(11):48–59. https://doi.org/10.18799/24131830/2020/11/2885. In Rus
10. Mytnikov A, Saqib M (2018) Development of advanced technology for transformer winding condition control based on pulsed method. Proceedings of the IV annual international scientific conference, fundamental and applied sciences: the main results of 2018. 12–13 December 2018, North Charleston, SC, USA, pp 14–17

11. Lavrinovich VA, Lavrinovich AV, Mytnikov AV (2018) Sposob kontrolya mekhanicheskogo sostoyaniya obmotok transformatora [Method for monitoring the mechanical state of transformer windings]. Patent RF № 2018116811/28 (026173); prioritet ustanovlen po date 04.05.2018 – RF Patent No. 2018116811/28 (026173); priority is set by the date 05/04/2018. In Rus

12. Lavrinovich VA, Lavrinovich AV, Mytnikov AV (2020) Eksperimental'noye issledovaniye kontrolya sostoyaniya obmotok vysokovol'tnykh transformatorov na osnove kommutatsionnykh impul'sov [Experimental study of monitoring the state of the windings of high-voltage transformers based on switching impulses]. Izvestiya Tomskogo politekhnicheskogo universiteta. Inzhiniring georesursov – Bulletin of the Tomsk Polytechnic University. Engineering of georesources 331(5):77–86. In Rus

Chapter 5
Development of a Schematic Diagram and a Probing Pulsed Generator Prototype

This chapter discusses the design stages of a special impulse generator for obtaining probing pulses. This generator is the main part of the diagnostics complex for nanosecond test control of transformer windings. A comparative analysis of various types of switches to provide the necessary parameters of the probing pulse is presented. A designed generator view and technical characteristics are given.

A voltage pulse generator (VPG) is the main technical unit of a diagnostic complex. The main technical requirements for it are presented in Table 5.1.

The use of a source of probing pulses with such parameters makes it possible to obtain a wider spectrum of generated frequencies and, accordingly, to increase the sensitivity of diagnostics to minor defects, such as displacement of turns, slight bulging of turns, unpressing, etc. The absence of a wave reflected from internal VPG elements increases the reliability of received information and greatly simplifies the diagnostic procedure of such a complex structure as a power transformer.

On the Russian market of high-voltage equipment, there are no VPGs that fully meet the requirements arising from their purpose in this diagnostic technology. Among those serially produced by the Russian industry, some types of VPGs do meet the requirements, but only in a few parameters.

Vvedensky [1], has proposed a generator circuit that allows one to generate single pulses at any load. The generator circuit is shown in Fig. 5.1.

The generator works as follows. From a constant voltage source U through a charging resistance R_1, a forming coaxial radio-frequency cable RF is charged. A feature of the circuit is that the cable sheath is potential. When the charging voltage reaches a set value, the key K is triggered, which can be uncontrolled, triggered at a fixed voltage value for this key, or controlled, triggered at different voltages when a start pulse is applied to it from a separate generator. After turning on the key К, waves begin to propagate in both directions from it. A wave equal to half of U propagates from the resistance R_2, which, having reached R_3, is reflected with an amplitude determined by the generator load resistance, and returns to R_2, being absorbed in it. Another wave propagates from the "Output" of the generator and, having reached R_2, is absorbed in it without reflection. Thus, at the "Output" of the generator, a pulse is

© The Author(s), under exclusive license to Springer Nature Switzerland AG 2022
V. Ya. Ushakov et al., *Transformer Condition Control*, Power Systems,
https://doi.org/10.1007/978-3-030-83198-1_5

Table 5.1 Basic technical requirements for a generator

1.	Probing pulse amplitude, V	Up to 300
2.	Pulse duration, ns	50–500
3.	Rise time, ns	10–15
4.	Pulse repetition period, s	20–30
5.	Application and storage at temperature from	−40 °C up to +45 °C

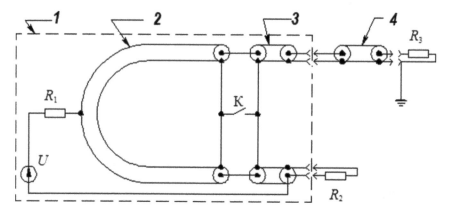

Fig. 5.1 Electrical circuit of a generator proposed by Yu.V. Vvedensky: U is a charging voltage source; 1 is common generator block; R_1 is a charging resistance; R_2 is a matching resistance equal to the wave impedance of the forming coaxial radio-frequency cable 2; R_3 is a load; K is a key, 2 is a forming a coaxial radio-frequency cable; 3 is a connecting coaxial radio-frequency cable located inside the generator; 4 is a transmitting coaxial radio-frequency cable 5–7 m long, connecting the generator to the load

formed with polarity reversed to the polarity of the charging voltage U, and duration equal to the travel time of the wave along the cable RK. Its amplitude depends on the ratio of wave resistance of an RF cable and the load resistance. Typical waveforms of pulses received from a generator at different loads are shown in Fig. 5.2. The pulses are obtained by simulating transients in the *MicroCAP* environment.

From the resulting oscillograms, it can be observed that for all types of load the duration and shape of a pulse on the load is the same: a single pulse without reflections. This is the advantage of such a generator circuit over the others in relation to the diagnosis of complex electrical circuits. Only the pulse amplitude changes depending on the ratio of a load and the internal resistance of a generator.

To change the pulse duration at the generator output, it is sufficient to change the length of the RF1 charging cable [1].

The most important element of a generator is a switch (K), the parameters of which significantly affect the front of an output pulse. During the development process, the following types of switches were investigated: mercury relay, mechanical slide switch of the BPK-2 M.01 type, push-button switch of the *DDR3 1066 SO-DIMM 1 Gb* type,

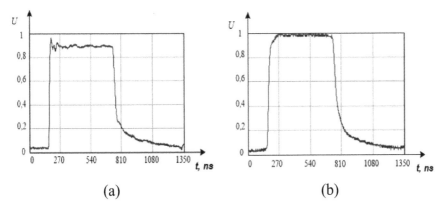

(a)   (b)

Fig. 5.2 Voltage pulses recorded at the matched load of the generator output: **a** mercury relay; **b** gas spark gap NENSHI 230-07

gas reed switch of the KEM type, vacuum reed switch of the MKA-52141-GrA type, low voltage arresters of the NENSHI-230-07 type.

Typical pulses generated by VPGs with such switches are shown in Fig. 5.2, where U is the voltage in relative units, t is the time in ns.

Visually, the pulses differ mainly in the front parameters. Some difference can also be observed in the flat part of pulses. This is due to the properties of switches applied, since other parameters of a generator, including the inductance of a switch connection, remained the same in all experiments. Comparison of pulses only by the rise time does not give a complete picture of the "filling" of a pulse with harmonics, which is important when probing transformer windings. First, measuring the rise time to compare the pulses with each other can produce a significant error, being dependent on the measurement method and the subjective characteristics of a measuring person (the "human factor" influence). Secondly, the comparison of pulses on the flat part is even more difficult for the reasons indicated above [2].

For an objective comparison of the received pulses with each other, we used the method of fast discrete Fourier analysis. The acceptability of the method is due to the fact that a discrete Fourier transform makes it possible to obtain the necessary results with speed and sufficient accuracy.

A pulse was decomposed into the components of a Fourier series using the standard *MathCAD* program. In this case, the duration of the pause between pulses was taken equal to twice the length of a formed pulse and remained constant in all cases, since when expanded in a Fourier series it affects the frequency spectrum and phase-frequency characteristic. In the described research, the duration of a pulse measured at half maximum was 675 ns, and the pause between pulses was 1350 ns. Before decomposition, all pulses were normalized in amplitude. The amplitude of each pulse was taken to be equal to one conventional unit. This made it possible to compare the amplitude-frequency characteristics of different switches, regardless of the amplitude of a generated pulse.

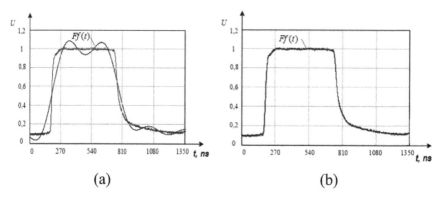

(a)   (b)

Fig. 5.3 Comparison of the original and reconstructed signals for a different number of harmonics with a switch of the NENSHI-230-07 gas discharge type: **a** N = 5; **b** N = 50 (N is the number of harmonics used in the inverse Fourier transform)

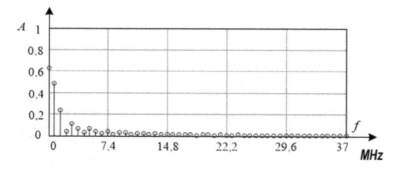

Fig. 5.4 Amplitude-frequency spectrum of a Vacuum Reed Switch MKA-52141-GrA

When using a Fourier series expansion, the question arose concerning the required number of harmonics in the expansion of an initial pulse. The following procedure was used. The pulse was decomposed into 50 harmonics, and then the original pulse was restored by reverse transformation using a limited number of harmonics [3]. The results are shown in Fig. 5.3.

In the images of reconstructed pulses displayed by Fig. 5.3, one can see that the minimum number of harmonics into which the initial pulse must be expanded is 50. With a smaller number of harmonics, there is a noticeable difference between the reconstructed pulses and the original one.

The efficiency of the method is confirmed by comparing the amplitude-frequency and phase-frequency characteristics of various types of switches (the amplitude-frequency characteristic in relative units is shown in Fig. 5.4, and the phase-frequency characteristic is shown in Fig. 5.5. Here, f is the frequency of the corresponding harmonic; $f = 0.74$ MHz is the fundamental frequency) [4].

The amplitude-frequency characteristics A for various types of switches can be defined as the doubled modulus of a discrete Fourier function:

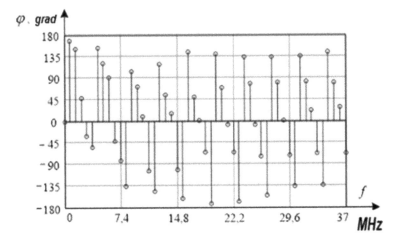

Fig. 5.5 Phase-frequency spectrum of a Vacuum Reed Switch

$$A_j = |F_j| \cdot 2, \tag{5.1}$$

where $j = 0\ldots50$ is the ordinal number of a harmonic, and A_0 is the amplitude of a constant component.

For other types of switches, the visual amplitude-frequency characteristics were but slightly different little from each other.

The phase-frequency characteristics φ for different types of switches can be determined using a function of the argument *arg* of a discrete Fourier function:

$$\varphi_j = \frac{\arg(F_j)}{\deg}, \tag{5.2}$$

where deg is a function that allows one to obtain phase-frequency characteristics in degrees, rather than in radians.

For other types of switches, the visual phase-frequency characteristics are also slightly different from each other. We consider it expedient to compare the spectra of phases and those of pulse amplitudes in the form of a relative deviation for the spectra of phases and spectra of amplitudes, and also compare the spectra of amplitudes in the form of spectral power density, i.e., the relative deviation of the energy contributions of each pulse harmonic to the total energy of the spectrum.

The relative deviation for the phase spectra of pulses $\Delta\varphi_j$ in relative units (r.u.) can be determined by the Eq. (5.3):

$$\Delta\varphi_j = \frac{\varphi_j - \varphi_{1j}}{\varphi_j}, \tag{5.3}$$

where φ_j and φ_{1j} are the phase spectra for two types of switches, respectively.

Relative deviation of the spectra of pulse amplitudes ΔAj in relative units equals to

$$\Delta A_j = \frac{A_j}{A_1}, \tag{5.4}$$

where A_j and A_{1j} are the amplitude spectra of two types of commutators.

The relative deviation of energy contributions for each harmonic of pulses to the total energy of the spectrum ΔW in relative units is equal to

$$\Delta W = \frac{W}{W_1}, \tag{5.5}$$

where: W is the contribution of each harmonic to the total energy of the switch spectrum of various types, with taking into account the energy of an individual harmonic, W1 is the contribution of each harmonic to the total energy of the spectrum of a switch of the gas gap type; $j = 0$–50 is the ordinal number of a harmonic; n = $2^{10} = 1024$ is the number of points. The peculiarity of a discrete Fourier transform is that the number of points must be a multiple of two.

An important step in the development of a diagnostic complex is to determine the criterion for choosing a switch.

The results of research showed that the main criterion for the acceptability of a certain switch for its use in generators for probing the mechanical condition of the windings of power transformers is the degree of filling with high-frequency components.

The suggested method for analyzing and comparing pulses at the output of a nanosecond generator with each other makes it possible to choose the most suitable switches in terms of filling the formed pulse with high-frequency components.

A vacuum reed switch of the MKA-52141-GrA type was selected as the main switch, and a gas reed switch of the KEM type or low voltage arresters of the *NENSHI*-230-07 type were selected as backup switches. A special switching board was developed, on which there was a BNC connector. Any of the listed switches can be quickly connected to this connector using a special case. For operational action, parallel to the main switch K, a mechanical slide switch of the BPK-2 M.01 type is connected with the output of the control button to the front panel of the switch being developed.

A separate power supply from a low-voltage transformer is provided to control the actuation of reed switches.

View of probing pulse generator is shown on Fig. 5.6.

All the generator elements are placed in a metal case, inside which there is a printed circuit board with soldered elements. Sections of forming RF cables RK1 are fixed on the chassis.

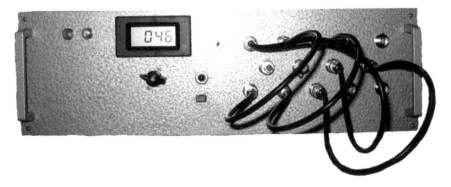

Fig. 5.6 General view of the generator

References

1. Vvedensky YuV (1959) Tiratronnyy generator nanosekundnykh impul'sov s universal'nym vkhodom [Thyratron generator of nanosecond pulses with a universal input]. Yu.V. Vvedensky. Izvestiya VUZov SSSR—Radiotekhnika—Izvestia of the USSR Universities.—Radio Engineering (2):249–251
2. Vasiljeva OV (2015) Programmnaya realizatsiya tsifrovoy obrabotki dannykh silovogo transformatora [Software implementation of digital processing of power transformer data]. Vasilieva, A.V. Lavrynovych // Programmnyye produkty i sistemy—Software products and systems (1):149–155 [358504-2015]
3. Vasiljeva OV (2016) An oscillograms processing algorithm of a high power transformer on the basis of experimental data. Vasiljeva OV, Budko AA, Lavrinovich AV. IOP Conference Series: Materials Science and Engineering. 124(1): Article number 012107. p 1–6 [595107–2016]
4. Lavrinovich AV (2015) Programmnyy kompleks dlya diagnostiki silovykh transformatorov [Software complex for diagnostics of power transformers]. Lavrinovich AV, Erfort AA. V sbornike trudov XX Vserossiyskaya nauchno-tekhnicheskaya konferentsiya studentov, aspirantov, molodykh uchenykh "Nauchnaya sessiya TUSUR–2015"—In the collection of works of the 20th All-Russian scientific and technical conference of students, graduate students, young scientists "Scientific session of TUSUR-2015".—Tomsk.—TUSUR.—Part 1. 250:255–258

Chapter 6
Single-Stage Defectography of Windings with Various Defect Degree Development

This chapter presents a new approach to the development of the pulse method for diagnosing transformer windings—a single-stage process of condition monitoring based on the use of a probing pulse only. The principle of control and analysis of diagnostic measurements are described on the example of a defect, the axial displacement of the winding turns. Possible prospects of this approach are discussed.

6.1 Displacement (Lodging) of Four Turns in the HV Winding of Phase a (Defect No. 1)

Spectral analysis was performed for a normogram and defect No. 1. Defect No. 1 is a mechanical displacement of 4 turns at the bottom of a winding. It is called the lodging of conductors and is implemented on the HV winding of phase "A". The description of first experimental results of this approach is given in [1, 2].

Figures 6.1 and 6.2 shows the oscillograms of pulses and the spectra ratio of a defect to the healthy winding (SR—Spectrum Ratio) of a transformer at the corresponding probe pulse durations of 520/260/110/60/20 ns (Figs. 6.3, 6.4, 6.5, 6.6, 6.7, 6.8, 6.9, 6.10, 6.11, 6.12, 6.13, 6.14 and 6.15).

6.2 Displacement (Lodging) of 8 Turns of the HV Winding of Phase a (Defect No. 2)

Control spectral analysis has been made for a normogram and defect No. 2, being a mechanical displacement of 8 turns at the bottom of a winding ("Lodging of conductors") of the HV phase "A".

Fig. 6.1 Oscillogram of a
520 ns pulse. Defect No. 1

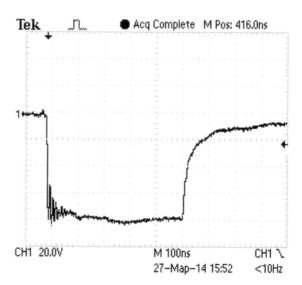

Fig. 6.2 Oscillogram of a
520 ns pulse for a healthy
winding

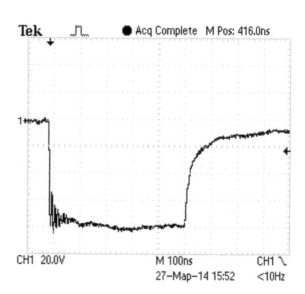

The figures below show only the spectra ratios of defective and healthy transformer
windings at the corresponding probe pulse durations of 520/260/110/60/20 ns. Oscil-
lograms of pulses look the same as in the cases of defectography of defect No. 1 (see
above) (Figs. 6.16, 6.17, 6.18, 6.19 and 6.20).

The obtained results are summarized in Table 6.1, which displays the maximum
values of resonance amplitudes at certain probe pulse durations for two defect
conditions.

Fig. 6.3 Frequency response ratio of a spectra 520 ns defect No. 1/healthy winding

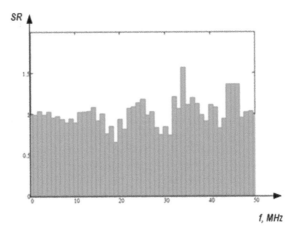

Fig. 6.4 Oscillogram of a 260 ns pulse for defect No. 1

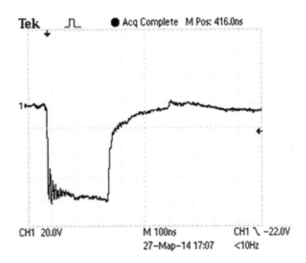

The data in Table 6.1 are also presented in the form of a graph, Fig. 6.21, which clearly shows the dependence of control efficiency on the duration of a probe pulse.

During the spectral analysis of signals obtained using the single-stage pulsed defectography of a transformer under investigation, and due to the obtained experimental results the following has been established:

- in principle, single-stage defectography is possible without registration of response signals from adjacent windings;
- This technology is not inferior in efficiency to the classical LVP method, and, due to its simplicity, surpasses the classical method in reducing the duration of a diagnostic process and the likelihood of making an error during diagnostics in industrial environment, as well as in the potential for monitoring the condition under operating voltage.

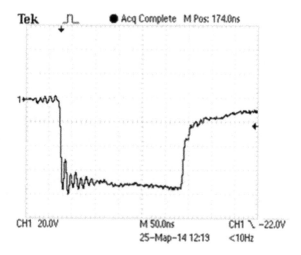

Fig. 6.5 Oscillogram of 260 ns pulse waveform for a healthy winding

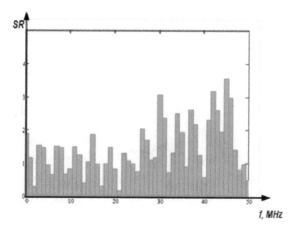

Fig. 6.6 Frequency response ratio of a spectra 260 ns defect No. 1/healthy winding

The technology of "single-stage defectography" allows detecting failures in a simpler and more convenient way under industrial conditions. It is possible to implement this method under operating voltage without disconnecting a transformer from a high-voltage network and unbussing by means of supplying a probe pulse to the winding under study through a voltage control sensor of the input.

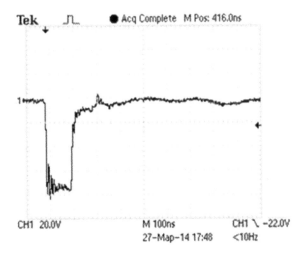

Fig. 6.7 Oscillogram of a 110 ns pulse for defect No. 1

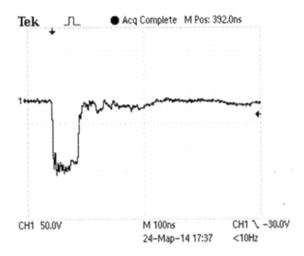

Fig. 6.8 Oscillogram of 110 ns pulse waveform for a healthy winding

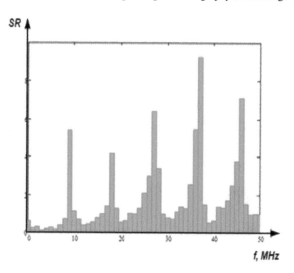

Fig. 6.9 Frequency response ratio of a spectra 110 ns defect No. 1/healthy winding

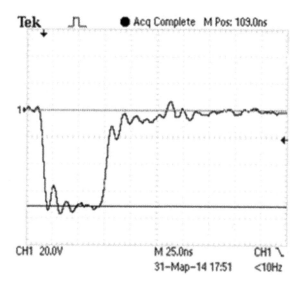

Fig. 6.10 Oscillogram of a 60 ns pulse for defect No. 1

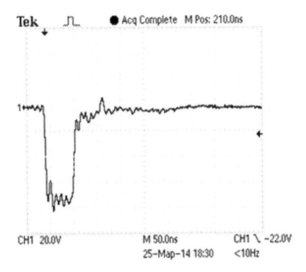

Fig. 6.11 60 ns pulse waveform for a healthy winding

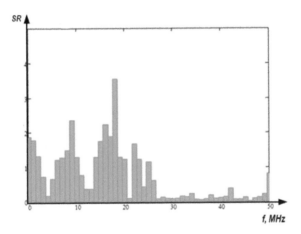

Fig. 6.12 Frequency response ratio of a spectra 60 ns defect No. 1/healthy winding

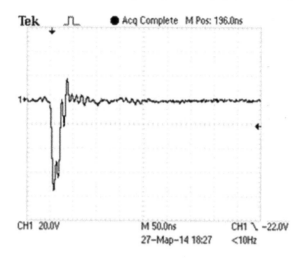

Fig. 6.13 Oscillogram of a 20 ns pulse for defect No. 1

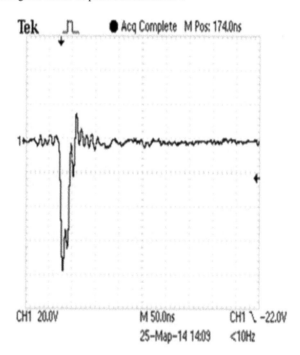

Fig. 6.14 Oscillogram of a 20 ns pulse for a healthy winding

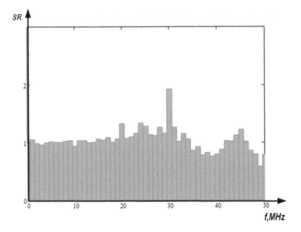

Fig. 6.15 Frequency response ratio of a spectra 20 ns defect No. 1/healthy winding

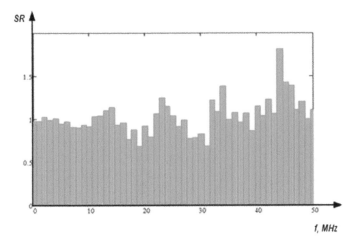

Fig. 6.16 Frequency response ratio of a 520 ns defect No. 2/healthy winding

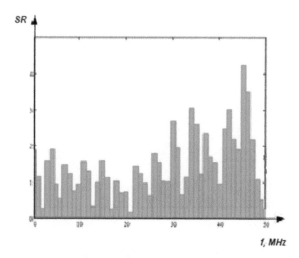

Fig. 6.17 Frequency response ratio of a 260 ns defect No. 2/healthy winding

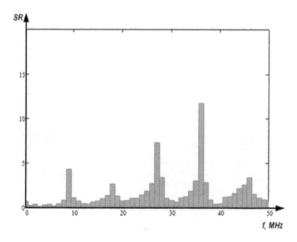

Fig. 6.18 Frequency response ratio of a 110 ns defect No. 2/healthy winding

Fig. 6.19 Frequency
response ratio of a 60 ns
spectra defect No. 2/healthy
winding

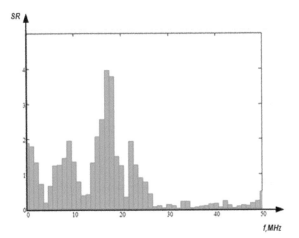

Fig. 6.20 Frequency
response ratio of a 20 ns
spectra defect No. 2/healthy
winding

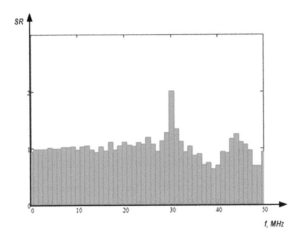

Table 6.1 Results of defect control for "lodging of conductors" in various degrees of development

Probe pulse duration, τ_{pulse}	Spectra ratio, Defect No. 1.	Spectra ratio, Defect No. 2
520	1.6	1.8
260	3.8	4.2
110	9	12
60	3.5	4
20	1.8	2

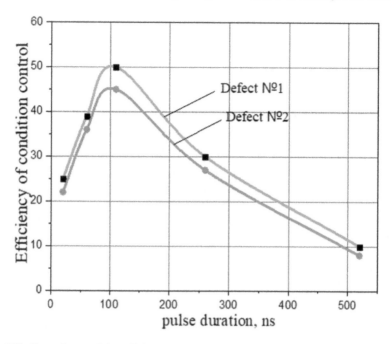

Fig. 6.21 Dependence of the efficiency of monitoring the winding condition in singe-stage defectography

References

1. Mytnikov A, Saqib M (2018) Development of advanced technology for transformer winding condition control based on pulsed method. Proceedings of the IV annual international scientific conference, fundamental and applied sciences: the main results of 2018—12–13 December 2018. North Charleston, SC, USA, pp 14–17
2. Mytnikov A, Lavrinovich A (2019) Further development of transformer winding condition control technology based on pulsed method. Proceedings of 14th International forum on strategic technology, IFOST-2019 October 14–17, 2019 Tomsk, Russia, pp 632–635

Conclusion

A big variety of possible defects developing in operating transformers requires the fast identification of dangerous fails during diagnostics. Currently, there is a common tendency of shift from the strategy of periodically conducted preventive work to preventive system, which only depends on the current condition state of the transformer. An effective preventive system ensures an increase in the reliability of equipment operation by reducing the number and duration of emergency and repair failures, timely withdrawal for repair and thereby extending the service life of the high voltage equipment. Prevention of emergency failure of power transformers allows obtaining such a significant economic effect that it becomes advisable to use even very complex means of measuring, processing data and issuing control results.

The organization of the optimal operation of transformers with a long service life requires the implementation of programs for a mass survey of the transformer fleet. Many countries have adopted multistep survey programs containing the minimum required set of methods for assessing the condition of transformers. In most cases, three-level transformer inspection programs are implemented: (1) routine control in the course of their operation, carried out by the services of power companies; (2) control by specialized organizations, which is more complex, requiring special equipment and, as a rule, disconnecting the transformer from the high voltage grid; (3) joint inspection with representatives of manufacturing plants according to a program developed individually for a given transformer with the determination of its operability.

The effectiveness of preventive maintenance according to the state is confirmed by the practice of mass surveys carried out recently. A full range of measures to assess the state of the transformer and determine its operability includes continuous and periodic monitoring of parameters in operation and detailed examinations in the disconnected state.

The lack of universal and unambiguous parameters for assessing the state of a transformer makes it necessary to use an optimally selected set of methods for its control, a comprehensive consideration of all the circumstances of the operation

V. Ya. Ushakov et al., *Transformer Condition Control*, Power Systems, https://doi.org/10.1007/978-3-030-83198-1

of 146 a transformer when deciding on its further work. It is especially difficult to decide when it is necessary to leave the transformer in operation, despite the presence of a defect. At the same time, the necessary restrictions on the operating mode are determined. Also, high qualification of technician staff is necessary to make a decision on where to repair the transformer—on working destination or at the transformer plant. As rule, such decisions require a participation of both side diagnostics group as well as representatives of the manufacturer.

The materials presented in this monograph will help to solve the abovementioned problems.

15753400R00166